ALL THE WORLD'S ANIMALS
HOOFED MAMMALS

ALL THE WORLD'S ANIMALS

HOOFED MAMMALS

TORSTAR BOOKS
New York · Toronto

CONTRIBUTORS

RJvA Rudi J. van Aarde
University of Pretoria
South Africa

RFWB Richard F. W. Barnes BSc PhD
Karisoke Research Centre
Ruhengeri, Rwanda
East Africa

GEB Gary E. Belovsky
University of Michigan
Ann Arbor, Michigan
USA

RAC Rosemary A. Cockerill PhD
Cambridge
England

DHMC David H. M. Cumming DPhil
Department of National Parks
and Wildlife Management
Causeway
Harare
Zimbabwe

GD G. Dubost PhD
Muséum National d'Histoire
Naturelle
Brunoy
France

HF Hans Frädrich PhD
Zoologischer Garten
Berlin
West Germany

WLF William L. Franklin MS PhD
Iowa State University
Ames, Iowa
USA

VG Valerius Geist PhD
University of Calgary
Calgary
Canada

SJGH Stephen J. G. Hall MA
University of Cambridge
England

HNH Hendrick N. Hoeck PhD
Universität Konstanz
Konstanz
West Germany

CJ Christine Janis MA PhD
Brown University
Rhode Island
USA

PJJ Peter J. Jarman BA PhD
University of New England
Armidale, NSW
Australia

JK Jonathan Kingdon PhD
University of Oxford
England

DWK David W. Kitchen PhD
Humboldt State University
Arcata, California
USA

KK Karl Kranz
Smithsonian Institution
Washington DC
USA

RML Richard M. Laws PhD FRS
British Antarctic Survey
Cambridge
England

WL Walter Leuthold PhD
Zurich
Switzerland

DFL Dale F. Lott
University of California
USA

SL Sandro Lovari PhD
Istituto di Zoologia
Parma
Italy

KMack Kathy Mackinnon MA DPhi
Bogor
Indonesia

MGM Martyn G. Murray PhD
University of Cambridge
England

PN Paul Newton BSc
University of Oxford
England

BPO'R Brian P. O'Regan BA MPhi
University of the Witwatersrand
Johannesburg
South Africa

NO-S Norman Owen-Smith PhD
University of the Witwatersrand
Johannesburg
South Africa

RAP Robin A. Pellew PhD
University of Cambridge
England

RJP Rory J. Putman BA DPhil
University of Southampton
Southampton
England

KR Katherine Ralls PhD
Smithsonian Institution
Washington DC
USA

DR Dan Rubinstein PhD
University of Princeton
Princeton, New Jersey
USA

MSP Mark Stanley Price DPhil
Office of the Adviser for
Conservation of the Environment
Oman

RU Rod Underwood MSc MIBiol BSc
University of Cambridge
England

PW Peter Wirtz PhD
Institut für Biologie
Freiburg
West Germany

ALL THE WORLD'S ANIMALS
HOOFED MAMMALS

TORSTAR BOOKS INC.
300 E. 42nd Street
New York, NY 10017

Project Editor: Graham Bateman
Editors: Peter Forbes, Bill MacKeith, Robert Peberdy
Art Editor: Jerry Burman
Picture Research: Linda Proud, Alison Renney
Production: Barry Baker
Design: Chris Munday

Originally planned and produced by:
Equinox (Oxford) Ltd
Mayfield House, 256 Banbury Road
Oxford, OX2 7DH, England

Editor
Dr David Macdonald
Animal Behaviour Research Group
University of Oxford
England

Artwork Panels
Priscilla Barrett

On the cover: Burchell's zebra
page 1: Mountain goat
pages 2–3: Burchell's zebra
pages 4–5: Elephant
pages 6–7: Domestic horse
pages 8–9: Reticulated giraffe

Printed in Belgium

Library of Congress Cataloging in Publication Data

Main entry under title:

Hoofed mammals.

(All the world's animals)
Bibliography: p.
Includes index.
1. Ungulata.   I. Series.
QL737.U4H66   1984   599.7   85–14116

ISBN 0–920269–72–9 (Series: All the World's Animals)
ISBN 0–920269–76–1 (Hoofed Mammals)

In conjunction with *All the World's Animals*
Torstar Books offers a 12-inch raised
relief world globe.
  For more information write to:
  Torstar Books Inc.
  300 E. 42nd Street
  New York, NY 10017

CONTENTS

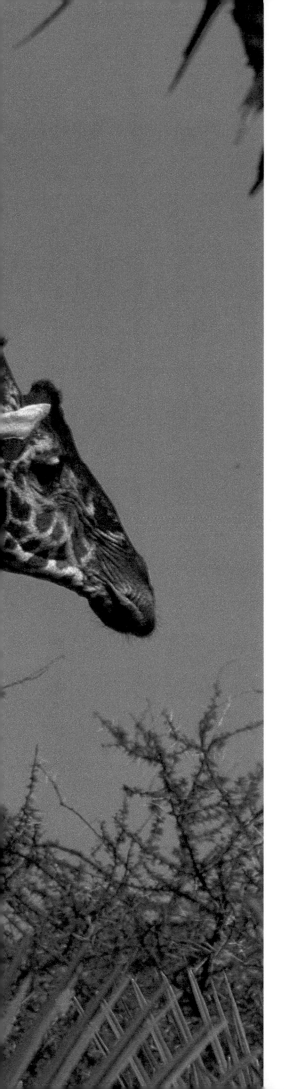

FOREWORD

The most spectacular and diverse array of land-dwelling animals, hoofed mammals range from the towering giraffe to the rabbit-sized mouse deer, from the ponderous rhino to the graceful, fleet-footed horse. Superbly adapted to every habitat, they occur from mountains to the African plains, from deserts to wetlands, and many, such as antelope, camels, sheep and cattle, have had enormous influence on human existence. Often elusive, always fascinating, these creatures and their lives are here portrayed in a vivid blend of words and pictures.

Hoofed Mammals is an ambitious, exciting journey into the kingdom of animals that will appeal alike to the professional and the schoolchild. The newest information and ideas are presented lucidly and entertainingly, but are far from being slight or superficial versions of the truth. The international panel of experts, whose work forms the heart of this book, have seen to it that the stories they tell are not only full of intrigue and incident but meet the highest scientific standards.

A superb collection of photographs and color drawings brings the text of this volume to life. With their aid, the reader will be an enthralled witness of the mighty struggles between American bison bulls, of the ritual confrontations of white rhinos and of the thundering journey of thousands of migrating wildebeest. The camera has captured, too, the lifestyle of creatures such as deer and pudu that shy away from human contact.

How this book is organized

Animal classification, even for the professional zoologist, can be a thorny problem and one on which there is scant agreement between experts. This volume has taken note of the views of many taxonomists but in general follows the classification of Corbet and Hill (see Bibliography) for the arrangement of families and orders.

This volume is structured at a number of levels. First, there is a general essay highlighting common features and main variations of the biology (particularly the body plan), ecology and behavior of the carnivores and their evolution. Second, essays on each family highlight topics of particular interest, but invariably include a distribution map, summary of species or species groupings, description of skull, dentition and unusual skeletal features of representative species and, in many cases, color artwork that enhances the text by illustrating representative species engaged in characteristic activities.

The main text of *Hoofed Mammals*, which describes individual species or groups of species, covers details of physical features, distribution, evolutionary history, diet and feeding behavior, as well as their social dynamics and spatial organisation, classification, conservation and their relationship with man.

Preceding the discussion of each species or group is a panel of text that provides basic data about size, life span and the like. A map shows its natural distribution, while a scale drawing compares the size of the species with that of a six-foot man. Where there are silhouettes of two animals, they are the largest and smallest representatives of the group. Where the panel covers a large group of species, the species listed as examples are those referred to in the text. For such large groups, the detailed descriptions of species are provided in a separate Table of Species. Unless otherwise stated, dimensions given are for both males and females. Where there is a difference in size between sexes, the scale drawings show males.

As you read these pages you will marvel as each story unfolds. But while discovering the fascination of these extraordinary creatures, you should also be fearful for them. Again and again, authors return to the need to conserve species threatened with extinction and by mismanagement. Of the 217 species described in these pages, about 74 species and subspecies are listed in the Appendices I through III of the Convention on International Trade in Endangered Species of Wild Flora and Fauna (CITES). The Red Data Book of the International Union for the Conservation of Nature and Natural Resources lists about 100 species at risk. In *Hoofed Mammals*, the following symbols are used to show the status accorded to species by IUCN at the time of going to press. ⒺE = Endangered—in danger of extinction unless causal factors are modified (these may include habitat destruction and direct exploitation by man). ⒱ = Vulnerable—likely to become endangered in the near future. ⓇR = Rare, but neither endangered nor vulnerable at present. ⒤ = Indeterminate—insufficient information available, but known to be in one of the above categories. ⑦ = Suspected, but not definitely known to fall into one of the above categories. The symbol ✴ indicates entire species, genera or families, in addition to those in the Red Data Book, that are listed by CITES. Some species and subspecies that have ⒺⓍ or probably have ⒺⓍ? become extinct in the past 100 years are also indicated.

PRIMITIVE UNGULATES

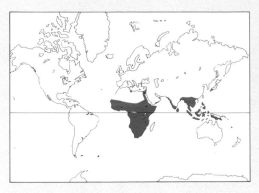

ORDERS: PROBOSCIDEA, HYRACOIDEA, TUBULIDENTATA†

Three orders: 6 genera; 14 species.

Elephants
Order: Proboscidea.
Two species: **African elephant** (*Loxodonta africana*), **Asian elephant** (*Elephas maximus*).

†A further order, the aquatic Sirenia, is considered to belong with the primitive ungulates, but is discussed only generally here.

Hyraxes
Order: Hyracoidea.
Eleven species in 3 genera.
Includes **Johnston's hyrax** (*Procavia johnstoni*). **Bruce's yellow-spotted hyrax** (*Heterohyrax brucei*).

Aardvark
Order: Tubulidentata.
One species: **aardvark** (*Orycteropus afer*).

THE relationship between the huge elephant and the tiny hyrax is not immediately obvious. Yet these creatures are primitive ungulates, and they, together with the aardvark, are thought to be more closely related to each other than to the other mammals. The aquatic sea cows of the order Sirenia (dugong and manatees) also share the same lineage as evidenced by the affinities between their ancestors.

The earliest ungulates, the Condylarthra, appeared in the early Paleocene (about 65 million years ago). They were the ancestors of the modern ungulates, the Perissodactyla (odd-toed ungulates) and Artiodactyla (even-toed ungulates). The Tubulidentata (now represented only by the aardvark) diverged early on during the Paleocene from the Condylarthra, and specialized in feeding on termites and ants.

Despite its resemblance to the other anteaters, the aardvark is not related to them at all. Its dentition is unique in that the front teeth are lacking and the peg-like molars and premolars are formed from columns of dentine. Fossil evidence suggests that members of this order spread from Africa to Europe and Asia during the late Miocene (about 8 million years ago), but three of the four genera recognized are now extinct.

Another Paleocene offshoot from the Condylarthra gave rise to the Paenungulata (the primitive ungulates or sub-ungulates) in Africa, during the continent's isolation. By the early Eocene (about 54 million years ago), the Paenungulata had separated into three distinct orders, the Hyracoidea (hyraxes), Sirenia (sea cows), and Proboscidea (trunked mammals, today represented only by the elephants).

The Hyracoidea proliferated about 40 million years ago, but spread no further than Africa and the eastern Mediterranean. Some of them became as large as tapirs, but today only the smaller forms remain. The decline of the Hyracoidea during the Miocene, 25 million years ago, coincided with the radiation of the Artiodactyla, against whom it is likely that they were unable to compete.

The Proboscidea were a very successful order that went through a period of rapid radiation in Africa and then spread across the globe except for Australia and Antarctica. The most obvious feature of the Proboscidea is of course their large size (see p14). Associated with this are their flattened soles, elongated limb bones, and the modifications to the head and associated structures. These include the evolution of the trunk and tusks, the elongated jaw and the specialized dentition, and the enlarged skull and shortened neck. In the middle Pleistocene the family Elephantidae enjoyed a period of very rapid evolution and radiation. Today only two species of elephant remain.

The three orders of the Paenungulata appear very different, but they share a few common anatomical and morphological features. None of them has a clavicle, and the primitive claws are more like nails than hooves. They all have four toes in the

► **The last outposts** of the hyraxes are the rocky outcrops or kopjes of East Africa. These are Johnston's hyrax and Bruce's yellow-spotted hyrax.

▼ **Evolution of the elephant.** Beginning with the small tapir-like *Mœnitherium* (1) in the early Oligocene (38 million years ago), proboscideans became a large, widespread group in the Pleistocene (2 million years ago). *Trilophodon* (2) was one of a family of long-jawed gamphotheres found in Eurasia, Africa, and North America from the Miocene to the Pleistocene (26–2 million years ago). *Platybelodon* (3) was a "shovel-tusked" gamphothere found in the late Miocene and Pliocene (about 12–7 million years ago) of Asia and North America. The Imperial mammoth (*Mammithera imperater*) (4), the largest ever proboscidean, flourished in the Pleistocene of Eurasia, Africa and North America. Unlike the earlier forms, it had high-crowned teeth like those of the modern African elephant (*Loxodonta africana*) (5).

forelegs and three toes in the hindlegs (elephants vary from this pattern); digits with short flattened nails (in hyraxes, the inner digit of the hindfoot bears a long curved claw); 20–22 ribs; similar anatomy of both placenta and womb; females have two teats between the forelegs (hyraxes have an additional two or four on the belly); the testes remain in the body cavity close to the kidneys. The orders are also related biochemically.

They all show developments of the grinding teeth and incisors, and they lack the other front teeth. The elephant's incisors have become its characteristic tusks, the dugong has a pair of upper incisor tusks, while the manatee has no front teeth at all. The hyrax has an enlarged upper incisor. All have transverse ridges upon their grinding teeth. The dugong has no premolars but large cusped molars. Elephants and manatees have unique dentition: in both species, the teeth form at the back of the jaw and are then pushed forward along the jaw. As they move forward, they are worn down with use. The elephant has six teeth in each jaw, only one being fully operational at a time, while the manatee has over 20, with about half a dozen in use at a time. Even more remarkable were the teeth of the Steller's sea cows: there were none at all. Instead it had horny grinding plates.

The Sirenia seem to have more in common with the Proboscidea than with the Hyracoidea. The Sirenia and Proboscidea may have diverged in the late Eocene from an ancestor whose fossils appear in the Fayum swamps of Egypt.   RFWB

ELEPHANTS

Order: Proboscidea
Family: Elephantidae.
Two species in 2 genera.
Distribution: Africa south of the Sahara;
Indian subcontinent, Indochina, Malaysia,
Indonesia, S China.

African elephant ⱽ ✱
Loxodonta africana
Distribution: Africa south of the Sahara.

Habitat: savanna grassland; forest.

Size: male head-body length 20–24.5ft
(6–7.5m), height 10.8ft (3.3m), weight up to
13,200lb (6,000kg); female head-body length
2ft (0.6m) shorter, weight 6,600lb (3,000kg).

Skin: sparsely endowed with hair; gray-black
when young, becoming pinkish white with age.

Gestation: 22 months.

Longevity: 60 years (more than 80 years in
captivity).

Subspecies: 2. **Savanna** or **Bush elephant**
(*L.a. africana*); E, C and S Africa. **Forest
elephant** (*L.a. cyclotis*); C and W Africa. The so-
called Cape elephant is regarded by some
authorities as a full subspecies. Once threatened
with extinction, it is now mainly restricted to
the Addo National Park in South Africa, where
it is rigorously protected and numbers are
increasing.

Asian elephant ᴱ ✱
Elephas maximus
Distribution: Indian subcontinent, Indochina,
Malaysia, Indonesia, S China.

Habitat: forest.

Size: head-body length 18–21ft (5.5–6.4m),
height 8.2–9.8ft (2.5–3m); weight up to
11,000lb (5,000kg).

Skin, gestation, and longevity: as for African
elephant.

Subspecies: 4. **Indian elephant** (*E.m.
bengalensis*), **Ceylon elephant** (*E.m. maximus*),
Sumatran elephant (*E.m. sumatrana*),
Malaysian elephant (*E.m. hirsutus*).

ᴱ Endangered.   ⱽ Vulnerable.   ✱ CITES listed.

ELEPHANTS have always been regarded
with awe and fascination, mainly because of their great size—they are the largest
living land mammals—and because of their
trunk and formidable tusks; but also
because of their longevity, their ability to
learn and remember and their adaptability
as working animals. For millennia, their
great strength has been exploited in agriculture and warfare and even today, notably in
the Indian subcontinent, they are still important economically and as cultural symbols. But a continuing demand for elephant
tusks, still the main source of commercial
ivory, has been largely responsible for a
drastic decline in elephant populations over
the past hundred years.

In the recent past the Asian elephant
ranged from Mesopotamia (now Iraq) in the
west, throughout Asia south of the
Himalayas to northern China. Today there
are fewer than 50,000 wild Asian elephants
remaining in isolated refuges in hilly or
mountainous parts of the Indian subcontinent, Sri Lanka, Indochina, Malaysia, Indonesia and southern China.

Of all elephants, the Savanna elephant is
the best understood—simply because it is
easier to study behavior in the open grasslands of eastern Africa than in the denser
forest habitats of the Forest and Asian
elephants.

Body size, the most conspicuous feature of
elephants, continues to increase throughout
life, so that the biggest elephant in a group is
also likely to be the oldest. The largest and
heaviest elephants alive today are African
Savanna bulls. The largest known speci-
men, killed in Angola in 1955 and now on
display in the Smithsonian Institution,
Washington DC, weighed 22,050lb
(10,000kg) and measured 13.1ft (4m) at
the shoulder. There have been several reports
in the past century of so-called Pygmy elephants, with an adult shoulder height of
less than 6.6ft (2m). It has been suggested
that these represent a separate species or
subspecies, but the current view is that they
are merely abnormally small individuals
which occasionally appear at random in
herds of normal-sized individuals.

The characteristic form of the skull, jaws,
teeth, tusks, ears and digestive system of
elephants are all part of the adaptive complex associated with the evolution of large
body size (see p14). The skull, jaws and
teeth form a specialized system for crushing
coarse plant material. The skull is dispropor-
tionately large compared with the size of the
brain and has evolved to support the trunk
and heavy dentition. It is, however, relatively light due to the presence in the upper
cranium of interlinked air-cells and cavities.
Asian elephants have two characteristic
dome-shaped protuberances above the eyes.

The tusks are elongated upper incisor
teeth. They first appear at the age of about 2
years and they grow throughout life so that
by the age of 60 a bull's tusks may average
132lb (60kg) each and a cow's 20lb (9kg)
each. In very old individuals they have been
known to reach 287lb (130kg) and attain a
length of 3.5m (7.7ft). Such massive "tus-
kers" have always been prime targets for
ivory and big game hunters, with the result
that few such specimens remain in the wild.
In general, the tusks of Asian bull elephants
are smaller than in their African counter-
parts, and among bull Ceylon elephants
they are formed in only 10 percent of
individuals. In the Forest elephant they are
thinner, more downward pointing and composed of even harder ivory than in the
Savanna subspecies. Elephant ivory is a
unique mixture of dentine, cartilaginous
material and calcium salts, and a transverse
section through a tusk shows a regular
diamond pattern, not seen in the tusks of
any other mammal. The tusks are mainly
used in feeding, for such purposes as prising
off the bark of trees or digging for roots, and
in social encounters, as an instrument of

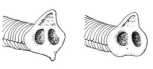

◄ **African and Asian elephants** compared. The Asian elephant (BOTTOM) is smaller than the African (TOP), has a convex back and much smaller ears. The trunk has two lips in the African elephant and one in the Asian. The small tusks of female Asian elephants are not visible beyond the lips.

▲ **The majesty** of the African elephant, seen here in Amboseli National Park, Kenya, with Mount Kilimanjaro towering behind it. The huge ears, which provide such a distinctive frontal appearance, function as radiators for the animal's bulky body, losing heat from their vast surfaces.

display, or they may be used as a weapon.

The upper lip and the nose have become elongated and muscularized in elephants to form a trunk. Unlike other herbivores, the elephant cannot reach the ground with its mouth, because its neck is too short. The option of evolving an elongated neck was not open to the early Proboscidea because of the weight of their heavy cranial and jaw structures. The trunk enables elephants to feed from the ground. It is also used for feeding from trees and shrubs, for breaking off branches and picking leaves, shoots and fruits. Though powerful enough to lift whole trees, the trunk, with the nostrils at its tip, is

also an acutely sensitive organ of smell and touch. Smell plays an important part in social contacts within a herd and in the detection of external threats. As to touch, the trunk's prehensile finger-like lips, endowed with fine sensory hairs, can pick up very small objects.

Further uses of the trunk include drinking, greeting, caressing and threatening, squirting water and throwing dust over its owner, and the forming and amplifying of vocalizations. Elephants drink by sucking water into their trunks then squirting it into their mouths; they also squirt water over their backs to cool themselves. At times of water shortage they sometimes spray themselves with the regurgitated water contents of their stomach. The trunk can also serve as a snorkel, enabling an individual to breathe if submerged, perhaps during a river crossing.

The elephant's large ears perform the same function as a car's radiator: they prevent overheating, always a danger in a large compact body with its relatively low rate of heat loss. They are well supplied with blood and can be fanned to increase the cooling flow of air over them. By evolving ears which substantially increase the body surface area, and so the rate of cooling, elephants have overcome one of the most important limitations to the evolution of large body size (see box). The ears are largest in the African Savanna elephant, which probably reflects the more open habitat in which this species lives. They are also more triangular than in the Forest elephant. Elephants have a keen sense of hearing and communicate extensively by means of vocalizations, particularly the forest-dwelling forms.

The massive body is supported by pillar-like legs with thick, heavy bones. The bone structure of the foot is intermediate between

that of man (plantigrade, where the heel rests on the ground) and that of the horse (digitigrade, where the heel is raised off the ground). The phalanges (fingers and toes) are embedded in a soft cushion of white elastic fibers enclosed within a fatty matrix. This enables the elephant to steal silently through the bush. The large surface area of the sole spreads the weight of their huge

▲ ► Versatile trunk. Really an extended, muscularized upper lip, the elephant's trunk is a sensitive all-purpose organ, giving it greater skill in handling food and other objects than any other ungulate. ABOVE An elephant having a mud-bath. The skin is sensitive, and dust- and mud-baths help to keep it free from parasites. ABOVE RIGHT An elephant drinking by squirting water, previously sucked into its trunk, into its mouth. BELOW RIGHT The trunk also allows them to feed on a wide range of food, from grasses to the leaves of trees.

▷ Wind-sniffers. OVERLEAF The trunk is an extremely sensitive sensory organ. These elephants are obviously picking up something interesting on the breeze.

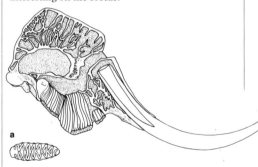

a

▲ Skull and teeth. The elephant's skull is massive, comprising 12–25 percent of its body weight. It would be even heavier if it were not for an extensive network of air-cells and cavities. The dental formula is I1/0, C0/0, P3/3, M3/3. The single upper incisor grows into the tusk and the molars (a) fall out at the front when worn down, being replaced from behind. Only one tooth on each side, above and below, is in use at any one time.

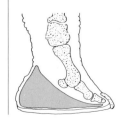

◄ The elephant's foot is broad and the digits are embedded in a fatty matrix (green). The huge weight of the animal is spread so well that it hardly leaves any track marks.

Evolution of Large Body Size

The elephant family were once highly successful, and during their peak they spread to all parts of the globe except Australia, New Zealand and Antarctica. Until the Pleistocene (about 2 million years ago), modern elephants occupied a range of habitats from desert to montane forest throughout Africa and southern Asia. This success was related to their most outstanding feature: the evolution of large body size.

In order to understand this, it is necessary to look at the early large herbivore community.

Different herbivore species in the same habitat avoid competing with each other for food by eating different plant groups (eg grasses, herbs, shrubs or trees), different species, or different parts (eg stem, leaf, fruit or flower) of the same species. The first large mammal herbivores in Africa were perissodactyls (eg the horses) which arose in the Eocene 54–38 million years ago) and dominated the large herbivore community until the coming of the ruminants (artiodactyls, eg the antelopes) in the early Oligocene (37 million years ago). It is likely that while each perissodactyl species ate a wide range of plants, taking the coarser parts, each ruminant species ate a narrower range, taking the softer parts.

The Proboscidea arose in the late Oligocene, followed by the Elephantidae in the Lower Miocene, so the ancestors of the elephants

arrived at a time when the large herbivore community had long been dominated by the perissodactyls and when the highly successful ruminants were continuing to evolve and to colonize new ecological niches. As non-ruminants, the Elephantidae were able to feed on plant foods which were too coarse for ruminants to live on. But this brought them into competition with the other non-ruminants there, the perissodactyls.

For a given digestive system, differences in metabolic rate enable a large animal to feed on less nutritious plant parts than can a smaller animal. Thus there was a strong selective pressure for the Elephantidae to increase their body size and so reduce the competition with perissodactyls. The most nutritious plant parts, such as leaf shoots and fruits, are produced only at certain seasons and even then may be sparse and widely scattered, but coarse plants are more abundantly distributed both in space and time. The elephants' evolutionary strategy thus enabled them to feed on plant parts which were not only abundant but available all the year round. In particular, it enabled them to feed on the woody parts of trees and shrubs. Thus they were able to tap a resource which other mammalian herbivores could neither reach nor digest. At the same time, they were able to eat rich plant parts (eg fruits) whenever these were available. This enabled elephants to thrive in a wide range of habitats.

and shrubs—twigs, branches and bark. They will also eat large quantities of flowers and fruits when these are available and they will dig for roots, especially after the first rains of the season. Asian elephants eat a similar range of foods, one of the most important being bamboo. Because of their large body size and rapid rate of throughput, elephants need large amounts of food: an adult requires about 330lb (150kg) of food a day. However, half the food leaves the body undigested.

Elephants cannot go for long without water, but because they are large, they can travel long distances each day between their water supplies and favorable feeding areas. They require 19–24gal (70–90 litres) of fluid each day, and at times of drought will dig holes in dry riverbeds with their trunks and tusks to find water. They also require shade, and during the hot season they spend the middle part of the day resting under trees to avoid overheating.

Elephants require a large home range in order to find enough food, water and shade in all seasons of the year. In one study, the home ranges of African elephants in a woodland and bushland habitat were found to average 290sq mi (750sq km) in an area of abundant food and water and 617sq mi (1,600sq km) in a more arid area. They respond very rapidly to sudden rainfall, often traveling long distances, up to 19mi (30km) to reach a spot where an isolated shower has fallen, in order to exploit the lush growth of grass which soon follows.

In both Asian and African forests, elephants often follow the same paths when moving from one place to another; over several generations this results in wide so-called "elephant roads" that can be found cutting through even the densest jungles.

Elephants are herd animals and display complex social behavior. Early hunters and naturalists spoke of a "herd bull" or "sire bull" which acted as a permanent leader and defender of the elephant herd. But several recent studies in East Africa have shown that bull and cow African elephants tend to live apart. Female elephants live in family units which are groups of closely related adult cows and their immature offspring. The adults are either sisters or mothers and daughters. A typical family unit may consist of two or three sisters and their offspring, or of one old cow and one or two adult daughters and their offspring. When the female offspring reach maturity, they stay with the family unit and start to breed. As the family unit grows in size, a subgroup of young adult cows gradually

bulk over such a wide area that on firm ground they leave hardly any track marks. Forest and Asian elephants usually have five toes on the forefoot and four toes on the hindfoot. The Savanna elephant typically has only four toes on the forefoot and three on the hindfoot.

Elephants walk at about 2.5–3.7mph (4–6km/h), but have been known to maintain double this speed for several hours. A charging or fleeing elephant can reach 25mph (40km/h), which means that over short distances they can easily outstrip a human sprinter.

The skin is 0.8–1.6in (2–4cm) thick and sparsely endowed with hair. Despite the thickness of the skin, it is highly sensitive and requires frequent bathing, massaging and powdering with dust to keep it free from parasites and diseases.

Elephants have a non-ruminant digestive system similar to that of horses. Microbial fermentation takes place in the cecum, which is an enlarged sac at the junction of the small and large intestines.

In the wet season, African Savanna elephants eat mainly grasses. They also eat small amounts of leaves from a wide range of trees and shrubs. After the rains have ended and the grasses have withered and died, they turn to feeding on the woody parts of trees

separates to form their own family unit. As a result, family units in the same area are often related.

The family unit is led by the oldest female, the matriarch. The social bonds between the members of the family are very strong. Cooperative behavior, particularly in the protection and guidance of the young, is frequently shown. When a hazard is detected (for example human scent) the group bunches with calves in the center and the matriarch facing the direction of the threat. If the matriarch decides to retreat, the unit runs in a very tight bunch. If she decides to confront the threat, the herd closely observe the outcome—which is normally the retreat of the threat. If one member is shot or wounded the rest of the group frequently comes to the aid of the stricken member even in the face of considerable danger—a behavior pattern which clearly plays into the hands of hunters.

If the matriarch is older than 50 years she may be reproductively inactive, for elephants, like humans, may survive after they are physically too old to reproduce. In other words, there is a menopause. This is a consequence of the elephant's longevity, which is in turn related to the evolution of large body size. Survival over a long lifespan both requires and facilitates the acquisition of considerable experience. By continuing to guide her family unit long after she is too old to breed, the matriarch can enhance the survival of her offspring by providing them with the benefits of her accumulated experience: her knowledge of their home range, of seasonal water sources and ephemeral food supplies, and of sources of danger and ways of avoiding them.

In contrast to their sisters, when the young male elephants reach puberty they leave the natal group. Adult bulls tend to live alone or in small temporary bull groups which are constantly changing in numbers and composition. Social bonds between bulls are weak and there seems to be little cooperative behavior.

Elephants, like other animals, communicate through sight, sound and smell. The most common vocalization is a growl emanating from the larynx (this is what hunters used to call the "tummy rumble"), a sound that carries for up to 0.6mi (1km). It may be used as a warning, or to maintain contact with other elephants. When feeding in dense bush, the members of a group monitor each other's positions by low growls. They vocalize less frequently when the bush is more open and the group members can see one another. The trunk is

used as a resonating chamber to amplify bellows or screams so as to convey a variety of emotions. The characteristic loud trumpeting of elephants is mainly used when they are excited, surprised, about to attack, or when an individual is widely separated from its herd.

Visual messages are conveyed by changes in posture and the position of the tail, head, ears and trunk. The ears and trunk evolved primarily for other purposes (as described earlier) but have a secondary value in communication. Elephants often touch each other using the trunk. Touch is especially important in mother-infant relations. The mother continually touches and guides her infant with her trunk. When elephants meet they often greet each other by touching the other's mouth with the tip of the trunk.

Most elephant populations show an annual reproductive cycle which corresponds to the seasonal availability of food and water. During the dry season, the population suffers a period of nutritional stress and cows cease to ovulate. When the rains break and the food supply improves, a period of one or two months of good feeding is needed to raise the females' body fat above the level necessary for ovulation. Thus females are in heat during the second half of the rainy season and the first months of the dry season.

Bulls of the Asian elephant have long been known to exhibit a condition during rutting called "musth"—a period of high male hormone (testosterone) levels, aggressive behavior, pronounced secretions from the temporal gland, and an increase in sexual activity. Musth usually lasts two or three months and tends to coincide with periods of high rainfall. Recently, African elephants have been found to show the same phenomenon. Another notable feature of bull elephants is that their testes remain within the body cavity throughout life and do not descend into a scrotum at puberty.

During the mating season, each female may be in heat for a few days only so the distribution of sexually receptive cows is constantly changing in space and time. Bulls travel long distances each day in order to monitor the changing reproductive status of cows in their home range. The bull who can travel longest and farthest during the mating season will find the greatest number of receptive females. The ability to travel fast at a low unit-energy expenditure is a further advantage of a large body. Having found a receptive cow, a bull will have to compete with other bulls for mating opportunities. Usually it is the largest bull who succeeds in copulating. It is this competition between males for females that has conferred evolutionary advantage on the size difference between males and females.

Following the long gestation, there is a long period of juvenile dependency. The infant suckles (with its mouth, not its trunk) from the paired breasts between the mother's forelegs for three or four years. Sexual maturity is reached at about 10 years of age, but it may be delayed for several years during drought or periods of high population density. Once she starts to breed, a cow may produce an infant every three or four years, although this period may also be extended when times are bad. The period of greatest female fecundity is between 25 and 45 years of age.

The long gestation period means that the infant is born nearly two years later, in the wet season, when conditions are optimal for its survival. In particular, abundant green food ensures that the mother will lactate successfully during the early months. During a birth, other cows may collect around the newborn elephant and so-called "midwives" may assist at the birth by removing the fetal membrane. Others may help the infant to its feet, and this marks the start of a joint family responsibility for the young of a group.

At birth, the African elephant weighs about·265lb (120kg) and the Asian about 220lb (100kg). The young elephant grows

▲ **Fighting among elephants** is usually restricted to play ritual, but on rare occasions it can be serious and even result in the death of one of the combatants if a tusk penetrates a vital organ. Fighting comprises a series of charges and head-to-head shoving matches which often involve wrestling with tusk and trunk.

Until about 10,000 years ago, man the hunter no doubt managed to kill some elephants as a source of meat, but it was not until modern man—the farmer—appeared that the first conflicts occurred (raiding elephants can destroy a crop overnight).

The first record of elephants being used as beasts of burden is found 5,500 years ago in the Indus Valley. Their natural characteristics of longevity, immense strength and placidity, and their ability to learn and remember, have been exploited up to the present day. Asian elephants have been kept in captivity continuously since that time, yet have never been completely domesticated. Although they are bred to a limited extent in captivity, there has always been a big demand for elephants captured from the wild, since an elephant has to be 10 years old before training can start and is not really useful until it is 20 years old. Also, bulls are less suited than cows to the captive way of life, which further reduces the chances of captive breeding. African elephants have had a more erratic record as beasts of burden. The most famous where those used by the Carthaginian leader Hannibal in the wars against the Romans, but recently

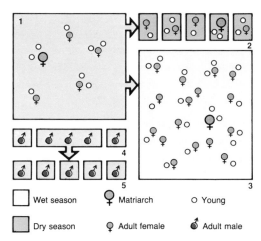

Wet season   ♀ Matriarch   ○ Young

Dry season   ♀ Adult female   ♂ Adult male

▲ ▼ **Group size in elephants.** (1) A typical family unit comprises closely related cows (with one dominant cow, the matriarch), and their offspring. (2) When food is scarce, the family groups tend to split up to forage. (3) In the wet season, family units may merge to form groups of 50 or more. Bulls leave the family group at puberty to join small, loose bull herds (4) or live alone (5). BELOW Mature bull elephants spend much of their time alone, moving between the female family units, searching for receptive cows.

rapidly, reaching a weight of 2,200lb (1,000kg) by the time it is 6 years old. The rate of growth decreases after about 15 years but growth continues throughout life. In addition, males experience a post-pubertal growth spurt between 20 and 30 years of age, which accounts for the size difference between them and females. Although the potential life span is about 60 years, half the wild elephants born die by the age of 15 years and only about one-fifth survive to reach 30 years of age.

The occurrence of elephant graveyards has been widely reported throughout the ages. However, there is no scientific evidence to support the theory that elephants migrate to specific sites to die. Large collections of bones can be accounted for in three ways: they may be the site of a mass slaughter of a herd by ivory traders or poachers or simply collections of bones washed to one site from a far wider area by flood waters or river systems; finally, they may be the site of the only water hole at a time of drought—elephants may have gathered there only to find food in such short supply, possibly due to bush fires, that they have starved to death.

Man's relationship with the elephants is beset by contradictions. On the one hand, elephants are valuable beasts of burden which need to be conserved; on the other hand, man is bringing them near to extermination in his thirst for land and ivory.

▲ **Ceremonial elephants.** Asian elephants still feature in ceremonies on the Indian subcontinent, as in this festival in Sri Lanka.

▼ **Working elephant.** Although declining in importance, the Asian elephant still has an economic role to play, particularly in the timber forests of Southeast Asia and the Indian subcontinent, where rough terrain and a lack of roads make it impossible to use tractors and lorries.

efforts have been made to use them in Zaire.

Elephants have always had another, more elevated, role in human affairs. Viewed with wonder and surrounded by mystique throughout its history, the elephant has found its way into the culture-myths and religions of all regions in which it has existed and even in modern times it features prominently in local religious symbolism and ceremonial.

In industrialized societies, elephants are held in such high regard as a popular spectacle, that for several centuries no circus or zoo has been without them. However, in the past, such establishments have depended almost entirely on the importation of wild stock and, as this stock diminishes, more attention will need to be paid to the breeding of elephants in captivity.

Although young elephants are often killed by lions, hyenas or tigers, the elephant's most dangerous enemy is man. In the 7th century BC, hunting for ivory caused the extinction of elephants in western Asia. In India, elephant numbers have steadily declined during the last millennium, as a result of sport hunting, ivory hunting and the spread of agriculture and pastoralism. In Africa, the Arab ivory trade, which started in the 17th century, caused a rapid decline of elephants in West Africa.

The colonial era, with the opening up of previously inaccessible areas and the introduction of modern technology, especially high-powered rifles, accelerated the decline of elephants. In Asia, this happened in the second half of the 19th century. In Africa, the destruction of elephants was highest between 1900 and 1910. Today, continuing deforestation and the spread of roads, farms and towns into former elephant habitats threaten both species by restricting their range, cutting off seasonal migration routes and bringing elephants into more frequent conflict with man. The human populations of the African countries south of the Sahara are doubling every 25–30 years, so intensifying the demand for land and the pressure on elephant habitats.

The worldwide economic recession of the early 1970s encouraged investors to switch to ivory as a wealth store. The sudden increase in the world price of ivory stimulated a wave of illegal hunting for ivory in Africa which is now causing a dramatic decline in elephant numbers. For instance, the elephant population of Kenya fell from 167,000 in 1973 to 60,000 in 1980. A recent wide-ranging survey estimates that 1.3 million elephants survived in Africa in 1980, but the fear is that this may be a considerable overestimate, bearing in mind the alarming rate at which elephants are still being killed (50,000–150,000 each year). In Asia, scarcely 50,000 elephants remain in the wild.

As their former habitats are destroyed and their numbers are cut by the greed for ivory, the only hope for the future conservation of elephants lies in the national parks. Ironically, in Africa, the existence of some national parks set up specifically to protect elephants is threatened by the elephants themselves. When a national park is created, mortality is reduced because poaching is controlled and access to water is guaranteed. The higher survival rates, especially among juveniles, causes an increase in elephant numbers within the park. At the same time, continuing human harassment outside drives elephants into the sanctuary of the park. Elephants have density-dependent mechanisms which regulate the population size. At high population densities there is an increase in the age of puberty, an increase in the interval between births, an earlier age of menopause and an increase in infant mortality. But under artificial conditions these mechanisms act too slowly. At

▲ Destructive elephants. In the dry season, elephants feed on trees, stripping and eating bark, demolishing whole trees to reach the leaves and twigs. In some parks, the elephant population is large enough to cause a rapid loss of the tree population and a conversion of the original habitat from woodland or bushland to grassland. The vegetation changes are probably accompanied by changes in the insect, bird and mammal populations, and possibly by changes in the soils and water table. Often the trends are exacerbated by fires which may be started by poachers or which may sweep in from outside the park. Fires prevent regeneration of woody species, kill trees already damaged by elephants, and destroy dry season browse, so that elephants are forced to concentrate on unburned areas, so causing even greater pressure on the vegetation.

high elephant densities, more trees and shrubs are killed by their feeding than can be replaced by natural regeneration. This results in a conversion of the habitat from woodland or bushland to grassland.

Fears that national parks would be irreversibly damaged by elephants caused a fierce controversy in the 1960s. Some conservationists argued that the elephant increase was due primarily to human activities and therefore humans should cull elephants to redress the ecological balance. They argued that while killing elephants is repugnant, especially in a national park, culling is necessary to prevent the loss of plants and animals that the park is supposed to conserve. Other conservationists argued that all animal and plant populations fluctuate naturally and that the elephant increase was just one phase of a natural cycle: soon numbers would go down, trees and shrubs would regenerate, and so there was no justification for culling. But in the last few years another factor has emerged: a lengthy drought has brought many elephants close to starvation, and in 1983 in Tsavo National Park, Kenya, some culling of dying elephants took place, leaving healthier herds.

African elephants could, in theory, be farmed, but the main drawback is their long reproductive cycle. However, in some overpopulated reserves, culled animals have already become a managed source of income—the ivory is sold at auction, the meat is dried and sold locally, the fat is turned into cooking oil, and the skin tanned and used for leather goods.

Since the late 1960s, the continent's political upheavals have also contributed to the elephant problem: the various conflicts and civil wars in Eritrea, Somalia, Sudan, Chad, Uganda, Zimbabwe, Mozambique, and Angola made semi-automatic and automatic weapons widely available throughout eastern and central Africa. These weapons are now falling into the hands of commercial poachers, making them the final arbiters of wildlife management in Africa. National park rangers, who are nearly always poorly armed and badly equipped, cannot contend with this new development. For the elephant, faced with modern weapons and the continuing loss of its habitat, the future is grim indeed. RFWB

HYRAXES

Order: Hyracoidea
Family: Procaviidae.
Eleven species in 3 genera.
Distribution: Africa and the Middle East.

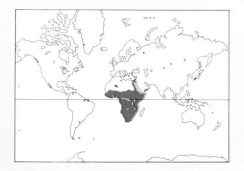

Rock hyraxes or dassies
Genus *Procavia*
Species: **Abyssinian hyrax** (*P. habessinica*),
Cape hyrax (*P. capensis*), **Johnston's hyrax**
(*P. johnstoni*), **Kaokoveld hyrax** (*P. welwitschii*),
Western hyrax (*P. ruficeps*).
Distribution: SW and NE Africa, Sinai to
Lebanon and SE Arabian Peninsula. Habitat:
rock boulders in vegetation zones from arid to
the alpine zone of Mt. Kenya 10,500–13,800ft
(3,200–4,200m); active in daytime. Size:
head-body length 17–21in (44–54cm); weight
4–12lb (1.8–5.4kg). Coat: Light to dark brown;
dorsal spot dark brown in Cape hyrax,
yellowish-orange in Western hyrax. Gestation:
210–240 days. Longevity: 9–12 years.

Bush hyraxes
Genus *Heterohyrax*
Species: **Ahaggar hyrax** (*H. antinae*), **Bruce's
yellow-spotted hyrax** (*H. brucei*), **Matadi
hyrax** (*H. chapini*).
Distribution: SW, SE to NE Africa. Habitat: rock
boulders and outcrops in different vegetation
zones in E Africa, sometimes in hollow trees;
active in daytime. Size: head-body length
12.5–18.5in (32–47cm); weight 2.9–5.3lb
1.3–2.4kg). Coat: light gray, underparts white,
dorsal spot yellow. Gestation: about 230 days.
Longevity: 10–12 years.

Tree hyraxes
Genus *Dendrohyrax*
Species: **Eastern tree hyrax** (*D. validus*),
Southern tree hyrax (*D. arboreus*), **Western
tree hyrax** (*D. dorsalis*).
Distribution: SE and E Africa (Southern tree
hyrax), W and C Africa (Western tree hyrax),
Kilimanjaro, Meru, Usambara, Zanzibar,
Pemba and Kenya coast (Eastern tree hyrax).
Habitat: evergreen forests up to about 12,000ft
(3,650m) (Southern tree hyrax among rock
boulders in Ruwenzori). Size: head-body length
12.5–24in (32–60cm); weight 3.7–9.9lb
(1.7–4.5kg). Longevity: greater than 10 years.
Coat: long, soft and dark brown, dorsal spot
light to dark yellow, from ¼–1½in long (Eastern
tree hyrax) to 1½–3in (Western tree hyrax).
Gestation: 220–240 days.

"THE high mountains are for the wild goats; the rocks are a refuge for the conies"—so runs the biblical characterization of the hyrax (Psalms 104:18). In Phoenician and Hebrew, hyraxes are known as *shaphan*, meaning "the hidden one." Some 3,000 years ago, Phoenician seamen explored the Mediterranean, sailing westward from their homeland on the coast of Syria. They found land where they saw many animals which they thought were hyraxes, and so they called the place "Ishaphan"—Island of the Hyrax. The Romans later modified the name to Hispania. But the animals were really rabbits, not hyraxes, and so the name Spain derives from a faulty observation!

The odd appearance of the hyrax has caused even further confusion. Their superficial similarity to rodents led Storr, in 1780, mistakenly to link them with guinea pigs of the genus *Cavia*, and he thus gave them the family name of Procaviidae or "before the guinea pigs." Later, the mistake was discovered and the group was given the equally misleading name of hyrax, which means "shrew mouse."

Hyraxes are small and solidly built, with a short rudimentary stump for a tail. Males and females are approximately the same size. The feet have rubbery pads containing numerous sweat glands, and are ill-equipped for digging. While the animal is running, the feet sweat, which greatly enhances its climbing ability. Species living in arid and warm zones have short fur, while tree hyraxes and the species in alpine areas have thick, soft fur. Hyraxes have long tactile hairs at intervals all over their bodies, probably for orientation in dark fissures and holes. They have a dorsal gland, surrounded by a light-colored circle of hairs which stiffen when the animal is excited. This circle is most conspicuous in the Western tree hyrax and least so in the Cape hyrax.

Early this century, fossil evidence showed that the hyraxes share many common features of primitive ungulates, especially elephants, and the related sirenians, and, as more recent research has indicated, also with the aardvark. Fossil beds in the Fayum, Egypt, show that 40 million years ago hyraxes were the most important medium-sized grazing and browsing ungulates during that time. Then there were at least six genera, ranging in size from that of contemporary hyraxes to that of a tapir. During the Miocene (about 25 million years ago), at the time of the first radiation of the bovids, hyrax populations were greatly reduced, surviving only among rocks and trees—

habitats that were not invaded by bovids.

Contemporary hyraxes retain primitive features, notably an inefficient feeding mechanism, which involves cropping with the molars instead of the incisors used by modern hoofed mammals, poor regulation of body temperature and short feet.

In the Pliocene (7–2 million years ago), hyraxes were both widespread and diverse: they radiated from southern Europe to China, and one fossil form, *Pliohyrax graecus*, was probably aquatic. Yet today they are confined to Africa and the Middle East.

The rock hyraxes have the widest geographical and altitudinal distribution, while the bush hyraxes are largely confined to the eastern parts of Africa. Both are dependent on the presence of suitable refuges in rocky outcrops (kopjes) and cliffs. As their name

▲ **Hyrax shows its teeth.** Hyraxes, like this Johnston's hyrax, can fight vigorously if cornered, biting savagely with their incisors. Curiously, these teeth are not much used in cropping, the molars being used instead, a relatively inefficient cropping method.

suggests, the tree hyraxes are found in arboreal habitats of Africa, but in the alpine areas of the Ruwenzori Mountains they are also rock dwellers. The Eastern tree hyrax might be the earliest type of forest-living tree hyrax, being a member of the primitive fauna and flora of the islands of Zanzibar and Pemba.

Hyraxes feed on a wide variety of plants. Rock hyraxes feed mainly on grass, which is a relatively coarse material, and therefore have hypsodont dentition (high crowns with relatively short roots), whereas the browsing bush hyraxes and tree hyraxes consume softer food and have a brachydont dentition (short crowns with relatively long roots).

Hyraxes do not ruminate. Their gut is complex, with three separate areas of microbial digestion, and their ability to digest fiber

efficiently is similar to that of ruminants. Their efficient kidneys allow them to exist on minimal moisture intake. In addition, they have a high capacity for concentrating urea and electrolytes, and excrete large amounts of undissolved calcium carbonate. As hyraxes have the habit of always urinating in the same place, crystallized calcium carbonate forms deposits which whiten the cliffs. These crystals were used as medicine by several South African tribes and by Europeans.

Hyraxes have a poor ability to regulate body temperature, and a low metabolic rate. Body temperature is maintained mainly by gregarious huddling, long periods of inactivity, basking and relatively short periods of activity. In summary, their physiology allows them to exist in very dry areas with food of poor quality, but they are dependent on shelter which provides them with an environment of relatively constant temperature and humidity.

Different species of hyraxes can co-exist in the same habitat (see box).

Groups of Bruce's yellow-spotted hyrax and Johnston's hyrax live on rock outcrops in the Serengeti National Park, Tanzania. Together, these two species are the most important resident herbivores of the kopjes. Their numbers depend upon the area of the kopje. The population density ranges for bush hyraxes from 8–21 animals per acre and for the rock hyraxes from 2–16 animals per acre. The group size varies between 5 and 34 for the former and for the latter between 2 and 26. Taking the two species together, this is comparable to the density of wildebeest in the long grass plains surrounding the kopjes. Among the different groups, the adult sex ratio is skewed in favor of females (1.5–3.2:1 for Bruce's yellow-spotted hyrax and 1.5–2:1 for Johnston's hyrax), while the sex ratio of newborns is 1:1.

The social organization varies in relation to living space. On kopjes smaller than 43,000sq ft (4,000sq m), both rock and bush hyraxes live in cohesive and stable family groups, consisting of 3–7 related adult females, one adult territorial male, dispersing males and the juveniles of both sexes. Larger kopjes may support several family groups, each occupying a traditional range. The territorial male repels all intruding males from an area largely encompassing the females' core area (average for bush hyraxes, 22,600sq ft; 2,100sq m; 27 animals, and for rock hyraxes 45,750sq ft; 4,250sq m; 4 animals).

The females' home ranges are not defended and may overlap. Rarely, an adult female will join a group, and such females are eventually incorporated into the female group. Females become receptive about once a year, and a peak in births seems to coincide with rainfall. Within a family group, the pregnant females all give birth within a period of about three weeks. The number of young per female bush hyrax

Kopje Cohabitants

The dense vegetation of the Serengeti kopjes supports two species of hyraxes—the gray-brown Bruce's yellow-spotted hyrax and the larger, dark brown Johnston's hyrax—living together in harmony. Whenever two or more closely related species live together permanently in a confined habitat, at least some of their basic needs like food and space resources must differ, otherwise one species will eventually exclude the other.

When bush and rock hyraxes occur together they live in close contact. In the early mornings they huddle together, after spending the night in the same holes. They use the same urinating and defecating places. Newborns are greeted and sniffed intensively by members of both species. The juveniles associate and form a nursery group; they play together with no apparent hindrance as play elements in both species are similar. Most of

their vocalizations are also similar, such as sounds used in threat, fear, alertness and contact situations. Such a close association has never been recorded between any other two mammal species except primates. However, bush and rock hyraxes do differ in key behavior patterns. Firstly, they do not interbreed, because their mating behavior is different and they have also different sex organ anatomy: the penis of the bush hyraxes being long and complex, with a thin appendage at the end, arising within a cup-like glans penis, and that of the rock hyrax being short and simple. Secondly, the male territorial call, which might function as a "keep out" sign, is also different, and, finally, the bush hyrax browses on leaves while the rock hyrax feeds mainly on grass. The latter is probably the main factor that allows both species to live together.

varies between 1 and 3 (mean 1.6) and in rock hyraxes between 1 and 4 (mean 2.4). The young are fully developed at birth, and suckling young of both species assume a strict teat order. Weaning occurs at 1–5 months and both sexes reach sexual maturity at about 16–17 months of age. Upon sexual maturity, females usually join the adult female group, while males disperse before they reach 30 months. Adult females live significantly longer than adult males.

▼ **Mating in hyraxes** is brief and vigorous. The penis anatomy varies between the three genera. In rock hyraxes it is short, simply built and elliptical in cross section; in tree hyraxes it is similarly built and slightly curved; in bush hyraxes, shown here, it is long and complex: on the end of the penis, and arising within a cup-like glans penis is a short, thin appendage, which has the penis opening. (1) The male presses the penis against the vagina. (2) Violent copulation, in which the male leaves the ground. (3) The female moves forward causing the male to withdraw.

There are four classes of mature male: territorial, peripheral, early and late dispersers. Territorial males are the most dominant. Their aggressive behavior towards other adult males escalates in the mating season, when the weight of their testes increases twenty-fold. These males monopolize receptive females and show a preference for copulating with females over 28 months of age. A territorial male monopolizes "his" female group all year round, and repels other males from sleeping holes, basking places and feeding areas. Males can fight to the death, although this is probably quite rare. While his group members feed, a territorial male will often stand guard on a high rock and be the first to call in case of danger. The males utter the territorial call all year round.

On small kopjes, peripheral males are those which are unable to settle, but which on large kopjes can occupy areas on the periphery of the territorial males' territories. They live solitarily, and the highest ranking among them takes over a female group when a territorial male disappears. These males show no seasonality in aggression but call only in the mating season. Most of their mating attempts and copulations are with females younger than 28 months.

The majority of juvenile males—the early dispersers—leave their birth sites at 16–24 months old, soon after reaching sexual maturity. The late dispersers leave a year later, but before they are 30 months old. Before leaving their birth sites, both early and late dispersers have ranges which overlap with their mothers' home ranges. They disperse in the mating season to become peripheral males. Almost no threat, submissive and fleeing behavior has been observed between territorial males and late dispersers.

Individuals of rock and bush hyraxes were observed to disperse over a distance of at least 1.2mi (2km). However, the further a dispersing animal has to travel across the open grass plains, where there is little cover and few hiding places, the greater are its chances of dying, either through predation or as a result of its inability to cope with temperature stress.

The most important predator of hyraxes is the Verreaux eagle, which feeds almost exclusively on them. Other predators are the Martial and Tawny eagles, leopards, lions, jackals, Spotted hyena and several snake species. External parasites such as ticks, lice, mites and fleas, and internal parasites such as nematodes and cestodes also probably play an important role in hyrax mortality.

In Kenya and Ethiopia it was found that rock and tree hyraxes might be an important reservoir for the parasitic disease leishmaniasis.

The Eastern tree hyrax is heavily hunted for its fur, in the forest belt around Mt. Kilimanjaro; 48 animals yield one rug. Because the forests are disappearing at an alarming rate in Africa, the tree hyraxes are probably the most endangered of all hyraxes.   HNH

▼ **Bush hyrax in a tree.** A young Bruce's yellow-spotted hyrax climbing. Despite their name, bush hyraxes generally inhabit the same rocky outcrops as rock hyraxes. They do, however, sometimes inhabit hollow trees.

AARDVARK

Order: Tubulidentata
Family: Orycteropodidae.
Orycteropus afer
Aardvark, antbear or **Earth pig**
Distribution: Africa south of the Sahara,
excluding deserts.

Habitat: mainly open woodland, scrub and
grassland; rarer in rain forest; avoids rocky hills.

Size: head-body length 41–51in (105–130cm);
tail length 18–25in (45–63cm); weight
88–143lb (40–65kg). Sexes same size.
Skin: pale yellowish gray with head and tail off-
white or buffy white (the gray to reddish brown
color often seen results from staining by soil,
which occurs while the animal is burrowing).
Females tend to be lighter in color.

Gestation: 7 months.

Longevity: up to 10 years in captivity.

Subspecies: 18 have been listed but most may
be invalid. There is insufficient knowledge
about the animal for firm conclusions to be
drawn.

▶ **Nocturnal pursuits of the aardvark.** Once it
has found a termite mound, an aardvark takes
up a sitting position and inserts its mouth and
nose, creating a V-shaped furrow. Although it
has not been directly observed, termites and
ants are presumably taken in by the long sticky
tongue. Feeding bouts, lasting 20 seconds to
seven minutes, are interrupted by short bursts
of active digging. Digging may continue until
the whole body enters the excavation. Termite
mounds are, however, not totally destroyed
during a single visit and an aardvark will feed
on a single mound on consecutive nights.

Few people have had the fortune of a close
encounter with one of Africa's most
bizarre and specialized mammals: the aardvark. This nocturnal, secretive termite- and
ant-eating mammal is the only living member of the order Tubulidentata. Thanks to its
elusiveness, it is one of the least known of all
living mammals.

Superficially, aardvark resemble pigs, in
possessing a tubular snout and long ears.
Their pale, yellowish body is arched and
covered with coarse hair which is short on
the face and on the tapering tail but long on
its powerful limbs. These are short, with four
digits on the front feet and five on the back.
The claws are long and spoon-shaped, with
sharp edges. The elongated head terminates
in a long, flexible snout and a blunt, pig-like
muzzle. A dense mat of hair surrounds the
nostrils and acts as a dust filter during
burrowing. The wall between the nasal slits
is equipped with a series of thick fleshy
processes which probably have sensory
capabilities. Aardvark have no incisor or
canine teeth and their continuously growing, open-rooted cheek teeth consist of two
upper and two lower premolars and three
upper and three lower molars in each jaw
half. The cheek teeth differ from those of
other mammals in that the dentine is not
surrounded by enamel but by cementum.

The aardvark has a sticky tongue—
round, thin and long—and well-developed
salivary glands. Its stomach has a muscular
pyloric area, which functions like a gizzard,
grinding up the food. Aardvark therefore do
not need to chew their food. Both males and
females have anal scent glands which emit a
strong-smelling yellowish secretion.

Aardvark feed predominantly on ants and
termites, with ants dominating the diet
during the dry season and termites during
the wet. When foraging, the aardvark keeps
its snout close to the ground, and its tubular
ears pointed forwards. This indicates that
smell and hearing play an important part in
locating food. Aardvark follow a zigzag
course when seeking food, and the same
route may be used on consecutive nights.
While foraging, they frequently pause to
explore their immediate surroundings, by
rapidly digging a V-shaped furrow with the
forefeet and by sniffing it intensively.

Little information is available on reproduction, but young are probably born just
before or during the rainy season, when
termites become more available. Only one
young, with a weight of approximately 4lb
(2kg), is born at a time. It will accompany
its mother when two weeks old and start
digging its own burrows at about six

months, but may stay with the mother until
the onset of the next mating season.

Aardvark are almost exclusively nocturnal and solitary. Two individuals tracked
by radio in the Transvaal (South Africa)
were more active during the first part of the
night (20.00–24.00 hours). They foraged
on both dark and bright moonlit nights but
took shelter in one of several burrow systems within their home range during spells
of adverse weather or when disturbed.
They foraged over distances varying from
1.2–3mi (2–5km) per night. Other studies
suggest that aardvark may range as far as
9mi (15km) during a 10-hour foraging
period—even as far as 19mi (30km) a night.

The only evidence of aardvark is normally
their burrows, of which there are three
types: the burrows made when looking for
food; larger temporary sites which may be
used for refuge and which occur throughout
the home range; and permanent refuge sites
where young are born. The latter often have
more than one entrance and are frequently
modified through digging, extend deeply
into the ground and comprise an extensive
burrow system up to 43ft (13m) long. An
aardvark can excavate a burrow very
quickly and, depending on the soil type, can
dig itself in within 5–20 minutes. Droppings
are deposited in shallow digs throughout
their range and covered with soil.

Aardvark share their habitat with a
variety of other termite- and ant-eating
animals, such as hyenas, jackals, vultures,
storks, geese, pangolin, Bat-eared fox and
aardwolf, but all of these also take other
prey, which reduces competition with the
aardvark. In being a specialized feeder,
aardvark are extremely vulnerable to
habitat changes. While intensive crop farming over vast areas may reduce aardvark
density, increased cattle herding, which
through trampling creates better conditions
for the termites, may increase their numbers. In general, however, until more is
known about the behavior and ecology of
the aardvark, little progress can be made in
formulating management policies.

Apart from aardvark flesh, which is said
to taste like that of pork, various parts of the
aardvark's body are prized. Its teeth are
worn on necklaces by the Margbetu,
Ayanda and Logo tribes of Zaire to prevent
illness and as a good-luck charm. Its bristly
hair is sometimes reduced to powder and,
when added to local beer, regarded as a
potent poison. It is also believed that the
harvest will be increased when aardvark
claws are put into baskets used to collect
flying termites for food. RJvA

THE HOOFED MAMMALS

ORDERS: PERISSODACTYLA AND ARTIODACTYLA

Thirteen families: 82 genera: 203 species.

Odd-toed Ungulates
Order: Perissodactyla—perissodactyls

Asses, Horses and Zebras
Family: Equidae—equids
Seven species in 1 genus.
Includes **African ass** (*Equus africanus*), **Domestic horse** (*E. caballus*), **Grevy's zebra** (*E. grevyi*), **Plains zebra** (*E. burchelli*).

Tapirs
Family: Tapiridae
Four species in 1 genus.
Includes **Brazilian tapir** (*Tapirus terrestris*).

Rhinoceroses
Family: Rhinocerotidae
Five species in 4 genera.
Includes **White rhino** (*Ceratotherium simum*).

Even-toed Ungulates
Order: Artiodactyla—artiodactyls

Suborder: Tylopoda—tylopods
Camels
Family: Camelidae—camelids
Six species in 3 genera.
Includes **Bactrian camel** (*Camelus bactrianus*), **guanaco** (*Lama guanicoë*).

"U NGULATE" is a general term given to all those groups of mammals which have substituted hooves for claws during their evolution. This character appears to follow from a commitment to a terrestrial lifestyle, with rapid locomotion and a herbivorous diet. Ungulates are relatively large animals, none less than 2.2lb (1kg) in body weight, and they comprise the majority of terrestrial mammals over 110lb (50kg). Living ungulates belong to two different orders which diverged from a common hoofed ancestor some 60 million years ago. Despite the superficial similarities between horses and cows, rhinos and hippos, tapirs and pigs, the former of each pair belongs to the Perissodactyla (odd-toed ungulates), and the latter to the Artiodactyla (even-toed ungulates), and the similarities between them have largely come about due to convergent evolution.

The Ungulate Body Plan
Despite the variety of bodily shapes and adornments, there is an underlying common theme to the two lineages of modern ungulates. From a 2¼-pound chevrotain to a three-ton hippopotamus, and from the ponderous rhinoceros to the graceful horse, ungulates generally have long-muzzled heads, held horizontally on the neck, barrel-shaped bodies carried on forelegs and hindlegs of roughly equal length, and small tails. Their skin is quite thick and tends to carry a coat of hairs (which may be air-filled for insulation) rather than soft fur. Compared to the primitive mammalian limb pattern, in which the foot has five digits, all of which are placed on the ground in locomotion, all ungulates have thickened, hard-edged keratinous hooves. There is a reduction in the number of toes, and a lengthening of the metapodials (the long bones in the fleshy parts of human hands and feet), with the resultant lifting of the foot, so that the animal is in effect balancing on the tips of its toes. The evolutionary climax of this type of limb—termed unguligrade—can be seen in the long slender limbs of horses and antelopes, where it bestows both speed and endurance.

As the names imply, odd- and even-toed ungulates differ in the type and degree of modification, and the number of toes. Even-toed ungulates (artiodactyls) have four or two weight-bearing toes on each foot, the weight-bearing axis of the limb passing between the third and fourth digits. All modern artiodactyls have lost the first digit. Odd-toed ungulates (perissodactyls) have a single toe or three toes together bearing the weight of the animal, with the axis of the limb passing through the middle (or single). All modern species have lost the first and fifth digits in the hindfoot, and the first digit in the forefoot. The metapodials are unfused and relatively short.

The earliest horses were three-toed ungulates, but since the Oligocene (35 million years ago) horses have borne their weight on the single third toe, with ligaments rather than a fleshy pad for support. In the fossil three-toed horses the second and fourth digits were much reduced, although they bore fully formed hooves and would have contacted the ground to provide additional support in extreme extension of the front foot in locomotion, such as in galloping and jumping; the metapodials were greatly elongated (although not fused together) to form a long slender limb. All living species of equids have reduced these side toes to proximal splint bones, and bear their entire weight at all times on an enlarged single hoof.

Senses
Ungulates have good but not exceptional hearing, small ears which can be rotated to detect the direction of a sound, an apparently very good sense of smell, and excellent eyesight. The eyes function well by day and

▼ **Built for running,** these gemsbok in Etosha National Park epitomize the grace of the antelopes, the most successful of the hoofed mammals.

Suborder: Suina—suoids

Pigs
Family: Suidae
Nine species in 5 genera.
Includes **Wild boar** (*Sus scrofa*).

Peccaries
Family: Tayassuidae
Three species in 2 genera.
Inludes **Collared peccary** (*Tayassu tajacu*).

Hippopotamuses
Family: Hippopotamidae
Two species in 2 genera.
Includes **hippopotamus** (*Hippopotamus amphibius*).

Suborder: Ruminantia—ruminants

Chevrotains
Family: Tragulidae—tragulids
Four species in 2 genera.
Includes **Water chevrotain** (*Hyemoschus aquaticus*).

Musk deer
Family: Moschidae—moschids
Three species in 1 genus.
Includes **Musk deer** (*Moschus moschiferus*).

Deer
Family: Cervidae—cervids
Thirty-four species in 14 genera.
Includes **Red deer** (*Cervus elaphus*), **Reindeer** (*Rangifer tarandus*).

Giraffe
Family: Giraffidae—giraffids
Two species in 2 genera.
Includes **giraffe** (*Giraffa camelopardalis*).

Cattle, antelope, sheep, goats
Family: Bovidae—bovids
One hundred and twenty-four species in 45 genera.
Includes **Pronghorn** (*Antilocapra americana*), **eland** (*Taurotragus oryx*), **impala** (*Aepyceros melampus*), **kob** (*Kobus kob*), **oryx** (*Oryx gazella*), **Thomson's gazelle** (*Gazella thomsonii*).

night, and give a fair degree of binocular vision, especially in open-country species, allowing the animals to judge distance and speed accurately. Their communication depends mainly upon sight and sound, with some use of scent marks, in open-country species; forest-dwelling artiodactyls are more dependent upon scent for social signalling. Perissodactyls lack the diversity of scent glands found on the feet and faces of many artiodactyls. They rely instead more on auditory communication, with frequent vocalizations and the production of a large variety of sounds, and they produce a much greater variety of facial expressions than do artiodacyls.

Food and Feeding
The evolution of the ungulate limb illustrates adaptation to a mobile open-country existence. Evolution of their teeth, skulls and digestive anatomy parallels changes in their locomotion. All ungulates are terrestrial herbivores, feeding on leaves, flowers, fruits or seeds of trees, herbs and grasses (although pigs and peccaries are characteristically more omnivorous, and may include roots, tubers and animals in the diet). With rare exceptions, all ungulates stand on the ground to feed; even the aquatic hippopotamus feeds on land. They cannot use forelimbs to manipulate food, as do primates and squirrels; nor do they fell trees to reach foliage, as do elephants or beavers. Their food has to be taken directly from the plant, or off the ground if it has fallen, with the lips, teeth and tongue, and these are appropriately modified.

Even though ungulates are herbivorous, not all plant food is of similar nutritive value. Plants tend to be abundant in carbohydrates such as sugars and starches, which are easily digested sources of energy. However, they are low in fat, and frequently low in protein, which is the source of the building blocks of amino acids essential for growth and repair of body tissues. The absence of abundant fat does not seem to constitute a critical problem for ungulates, and many have lost the gall bladder, which in other mammals is the source of bile salts that emulsify and break down fats.

But obtaining sufficient protein in the diet is a critical matter for all herbivores. This is particularly true for the smaller ungulates, which have relatively greater requirements and higher turnover of nutrients, and small antelopes have very occasionally been observed to catch and kill birds as an additional source of protein. Pigs will also consume carrion.

The most abundant source of vegetable

protein is in seeds, but these are small and widely dispersed. The more easily available sources of vegetation, such as leaves, are composed primarily of carbohydrates, especially when they are mature. The carbohydrates in vegetation are available in two forms; in the soluble cell contents, and in the fibrous cell wall casing of cellulose, which is indigestible to most mammals.

Ungulate dentition is adapted to grinding so as to mechanically disrupt the cell wall to release the digestible contents. The back of an ungulate's mouth functions like a mill, with large flat square molars which reduce plant matter to fine particles. In conjunction with this, the jaw musculature and the configuration of the jaw joint are modified so that the lower jaw can be moved across the upper with a sweeping transverse grinding motion, in contrast to the more up-and-down motion in other mammals that simply cuts and pulps the food. The high-crowned (hypsodont) cheek teeth are made to last a lifetime of continual abrasion. In ungulates there is typically a gap between the milling molars and the plucking incisors. Whether or not canine teeth are retained depends upon their use as weapons; they appear to have no feeding function in ungulates,

although in ruminants (members of the artiodactyl suborder Ruminantia) the lower canines are retained and modified to form part of the lower incisor row.

Artiodactyls such as pigs and peccaries, which select only non-fibrous vegetation such as fruit and roots, do not digest the cellulose content of vegetation, and have a digestive system which resembles that of other mammals. However, other ungulates have a more fibrous diet so must be able to digest the large quantities of cellulose that they must ingest along with the more easily digestible parts of the plant. To achieve this the ingested food is fermented by bacteria somewhere along the digestive tract, transforming it into products which can then be absorbed and utilized.

There is a critical difference between the complex digestive systems of perissodactyls and those suborders of artiodactyls which eat fibrous vegetation (Ruminantia and Tylopoda). The "ruminant" artiodactyls have their fermentation chamber containing microorganisms situated within a complex multi-chambered stomach. The ruminant itself can digest both the continually multiplying microorganisms that overspill into the rest of the digestive tract and the products of

THE UNGULATE BODY PLAN

▶ **Teeth.** Primitive herbivorous mammals have molars with separate cusps (bunodont), designed to pulp and crush relatively soft food. This type is seen in pigs (**a**). Fibrous vegetation is tough and ungulates have developed modifications of the bunodont pattern. In perissodactyls, such as the rhinoceros (**b**), shearing edges (lophs) have formed by a coalescing of the cusps to form two crosswise lophs and one lengthwise (lophodont). In horses (**c**) the lophs are very complex and folded (hypsodont). In ruminant artiodactyls, such as the ox (**d**), the cusps take on a crescent shape (selenodont).

▼ **Jaws.** The different modes of feeding of perissodactyls and ruminant artiodactyls are reflected in the size of the jaw and musculature. Non-ruminant grazers, like the horse (**a**), have to consume large quantities of tough fibrous food and the lower jaw is very deep and the masseter muscle, primarily used in closing the jaws, is very large. Ruminants, like the giraffe (**b**), spend much of their time chewing the already half-digested cud and the lower law and masseter muscle are much less pronounced.

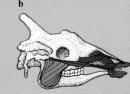

▲ **Mode of feeding.** Most perissodactyls (with the exception of some rhinos) retain both sets of incisors and use the upper lip extensively in feeding, like the horse (**a**). Ruminant artiodactyls, like the giraffe (**b**), have lost the upper incisors and make extensive use of a prehensile tongue rather than the upper lip. The resulting differences in facial musculature mean that perissodactyls have a much greater variety of facial expressions, used to communicate with each other, than do artiodactyls.

Hoofed Mammals' Feet

In the hoofed mammals the primitive mammalian foot (**a**) has been modified in various ways. The toes are reduced in number, and the long bones (metapodials) much extended. The foot is held lifted with only the tips of the toes on the ground (unguligrade). The joint surfaces are restricted so that the limbs cannot be rotated or moved in or out of the body to any great extent—the

prime movement is thus fore and aft, which facilitates fast running at the expense of climbing and digging. In the generalized ungulate feet (**b–e**), one or two digits are lost, the metapodials somewhat elongated, the tarsal bones are more ordered: (**b**) tapir, (**c**) pig, (**d**) peccary, (**e**) chevrotain. Rhinos (**f**) and hippos (**g**) have feet specialized for weight bearing (graviportal), with short digits and a spreading foot in which the side toes touch the ground when standing. In camels (**h**) the metapodials are long and fused for most of their length into a single bone. The most drastic modifications occur in the hoofed mammals adapted to fast running (cursorial). In the horse (**i**) the metapodials are totally fused and the digits are reduced to one (digits 2 and 4 are retained as vestigial splint bones). Deer (**j**) retain the side toes, and the metapodials are only partly fused. In the pronghorn (**k**) the fused metapodials are longest.

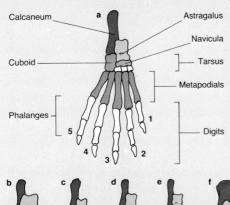

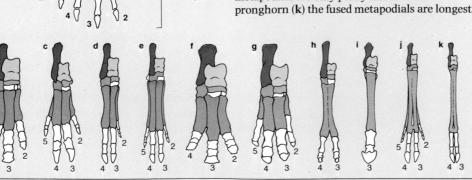

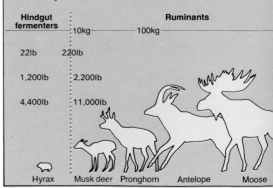

▼ Digestive systems. Ungulates have evolved two very different systems for dealing with the relatively indigestible cellulose in their highly fibrous food: hindgut fermentation (**a**) and rumination (**b**). In the hindgut fermenters (perissodactyls) food is completely digested in the stomach, and passes to the large intestine and cecum, where microorganisms ferment the ingested cellulose. Ruminants have a more complex digestive system and retain food in the gut for much longer. Food passes initially to the first stomach chamber (rumen) where it is fermented by microorganisms, and is regurgitated to be chewed and mixed with saliva. It then passes back to the second chamber (reticulum), bypassing the rumen. Bacteria spill over with the food and accompany it through the third (omasum) and fourth (abomasum) stomachs. Digestion is completed in the abomasum and nutrients are absorbed in the small intestine. Some additional fermentation and absorption occur in the cecum.

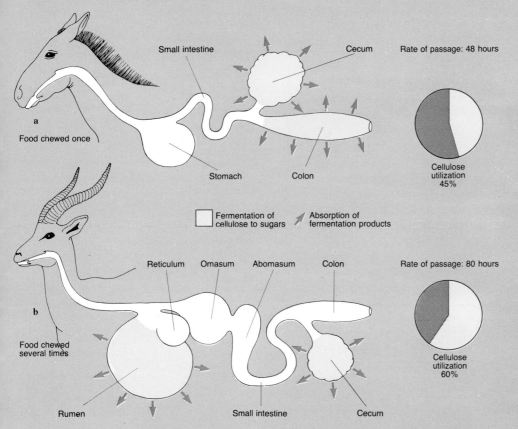

▼ Consequences of body size. Ruminants must pay a price for their highly efficient digestive system: it takes a long time for food to pass through their gut. In large herbivores, this price outweighs the advantages of efficient digestion. This is because large animals require absolutely more food per day than small ones, and they are therefore forced to accept low-quality food which can be gathered in large quantity. It takes so long to process low-quality food in a ruminant system that a greater net intake of nutrients is achieved with a simple gut and fast throughput. Hippos are the largest ruminants and appear to be an exception; however, they do not chew the cud and they have a fairly fast passage through the digestive system. Small animals require more food *per pound of body-weight* than large ones. Very small herbivores can select a high-quality diet which can be digested easily without time-consuming rumination. This is why no ruminants are less than 11lb (5kg) in weight.

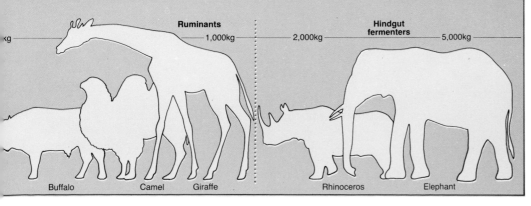

their fermentation of cellulose. The complex stomach allows food to be differentiated, digested food passing through a sieve-like structure to the posterior "true" stomach and the intestines, whereas undigested food is set aside to ferment, and is regurgitated to the mouth to be chewed again. This act is called rumination, familiar in domestic ruminants as "chewing the cud." This fermentation system is very efficient, making maximum use of the available cellulose, but is limited in that the food is retained for a very long time (up to four days), so that there is a lower limit to the amount of food that can be eaten and processed in a given period by a ruminant than a non-ruminant.

In contrast, in perissodactyls the hindgut areas of the cecum and colon (large intestine) are the site of fermentation. Although the processes of fermentation are biochemically identical to the ruminant process, the digestion of cellulose is less efficient, as the food is only retained for about half the length of time. However, this does mean that a greater quantity of food can be consumed per day, as the turnover rate is greater.

When comparing these ruminant and hindgut systems of fermentation, it is apparent that the latter is less efficient in the utilization of young, short herbage, which is high in protein and can yield all the requisite nutrients in small quantities. These can be utilized more efficiently by a ruminant. However, hindgut fermenters are at an advantage where food is of limited quality and high in fiber, thus necessitating a high intake to obtain sufficient protein, provided that this food is not also limited in quantity. Ruminants, on the other hand, are at an advantage in environments where food is of limited quantity, but where the quality is relatively high, for example desert inhabitants such as the oryx and camel, or arctic tundra inhabitants such as reindeer or Musk ox. In habitats such as the tropical savannas of Africa, where both types of animal co-exist, there is a partitioning of cropping, zebras (perissodactyls) for example eating the poorer-quality old foliage at the top of the grass stand, and gazelles and wildebeest (artiodactyls) eating the higher-quality new foliage uncovered by the zebras.

Absorption of the products of protein digestion is comparable in ruminants and hindgut fermenters, but ruminants can additionally recycle urea, a nitrogen-rich waste product that is normally excreted in the urine, using it to feed the microorganisms which they later digest. An important result of this difference is that perissodactyls such as asses which inhabit desert areas, need to drink daily to produce sufficient water to balance the urea in the urine, whereas ruminants, such as the oryx and camel, need to drink only occasionally.

Perissodactyls can make maximum use of fruit in their diet, as its essential nutrients, particularly sugars, are absorbed before the region of fermentation is reached. Ruminants ferment fruit in the fore-stomach and hence lose much of its nutritional value. The only ruminants that gain benefit from eating fruits are very small, such as chevrotains and duikers, which eat little fibrous food and have a small rumen.

There are more species of ruminants than there are of the hindgut-fermenting perissodactyls. The reason for this seems to be that the ability of ruminants to recycle nitrogen and to digest the protein-rich bacteria frees them from having to obtain all their essential amino acids from their diet, which means that they can afford to specialize on a narrower range of plant species in the environment. This in turn means that ruminants can subdivide the available niches in a habitat in a finer fashion.

Many living ungulates, such as tapirs, giraffes and many species of deer, are browsers, eating the leaves of dicotyledonous plants such as trees and shrubs. This was probably the original diet of all perissodactyls and ruminant artiodactyls.

But in the Miocene (about 20 million years ago) the monocotyledonous grasses emerged as abundant land plants, which comprise an excellent and abundant source of nutritious, although cellulose-rich, vegetation. The main difference between grasses and dicotyledonous plants as a dietary source lies in the structure and distribution patterns of the plants.

Grasses grow in great expanses, but their nutritive leaves are at the base of the plant, and valuable feeding time can be wasted in searching for them through a stand of fibrous stems of low nutritional value. Leaves on trees and shrubs are on the perimeter of the plant and are more easily accessible, but the actual plants themselves are spread further apart and must be located. Moreover, a bitten grass leaf can grow from the base to replace itself rapidly, but a tree or shrub grows from the apex, and cannot rapidly regenerate. As a consequence, trees and shrubs tend to arm themselves with thorns and unpalatable chemicals in the leaves to discourage destruction, which may make them hard to feed on. In contrast, grasses do not seem to "mind" being eaten so much, and are not similarly

defended. However, grass is only a really good nutritional source for certain parts of the year, in the growing season, whereas trees and shrubs provide a more predictable source of nutrients that are available the whole year round.

Most ungulate species have opted for becoming either predominantly browsers or predominantly grazers, and the most successful and abundant have been the grazers, the bovids and the horses, which radiated with the Miocene grasslands, evolving the mouth parts and digestion to cope with the high cellulose and high silica content of grasses, and the agility to avoid predators which were attendant risks of feeding in the exposed conditions where grasses grow.

▲ **Eocene ungulates.** During the Eocene (54–38 million years ago), the first hoofed mammals, which had evolved in the preceding Mesozoic and Paleocene, rapidly evolved to fill a wide variety of environmental niches. The perissodactyls (odd-toed ungulates) and the artiodactyls (even-toed ungulates), appeared simultaneously in Europe and North America. Odd-toed ungulates predominated in these early forms. (**1**) *Uintathere*, a large grotesque herbivore. (**2**) *Dolichorhinus*, a titanothere from the late Eocene. (**3**) *Eohippus*, the "Dawn horse" from the lower Eocene. (**4**) *Hyrachyus*, a small "running" rhinoceros. (**5**) *Amynodentopsis*, a semi-aquatic rhinoceros. (**6**) *Phenacodus*, a primitive hoofed herbivore. (**7**) *Meniscotherium*, a small browsing herbivore.

▼ **The evolution of hoofed mammals.**

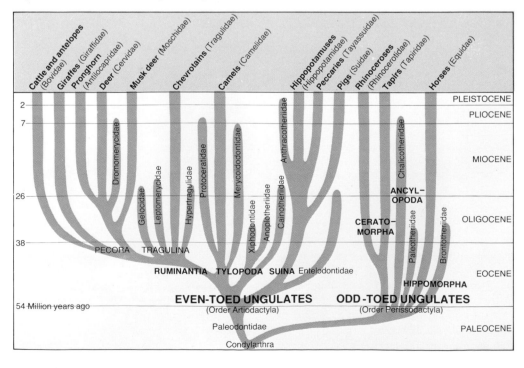

Evolution

Perissodactyls and artiodactyls first radiated in the early Eocene in the Northern Hemisphere 54 million years ago, at a time when Africa was isolated from Eurasia, and North America was isolated from South America, although a limited amount of interchange of animals was possible between the northern continents. The climate was warmer than today, and less differentiated latitudinally, with tropical forest reaching northwards to within the Arctic circle.

Eocene perissodactyls were a wide range of body sizes, ranging from about 11lb (5kg) to about 2,200lb (1,000kg), and were the first ungulates to adopt a diet of relatively fibrous vegetation (although all of them were browsers).

In North America the small Eocene equids and tapirs were selective browsers; with tapirs being more exclusively leaf-eaters, but equids handling a greater amount of fiber. Rhinos and chalicotheres were medium-sized browsers. The brontotheres were large animals, earlier forms being smaller and hornless, but later ones attaining the size of a White rhino and developing Y-shaped bony horns on the nose. Their teeth suggest a mixed diet of leaves and fruit. In Europe the brontotheres were absent, and their fruit specialist niche was probably taken by the endemic equid-related paleotheres, some of which sported a tapir-like proboscis. The only tapirs in the European faunas were the large and aberrant lophiodonts, which mimicked the rhinos of North America and Asia, and the European lineage of equids (which were all extinct by the end of the Eocene) diverged to fill the niches taken by both equids and tapirs in North America. In contrast, equids and paleotheres were largely absent from Asia, where the tapirs experienced the peak of their diversity during the Eocene, producing rabbit-sized and gazelle-like forms as well as more normal varieties.

In contrast to this blossoming of perissodactyl diversity, Eocene artiodactyls were all small (under 11lb), and were omnivorous or fruit-eating. This primary omnivorous tendency is retained today in the pig-like artiodactyls. However, there was never an equivalent radiation of omnivorous perissodactyls.

At the end of the Eocene (38 million years ago), the global climate changed dramatically, resulting in a seasonal climate in the northern continents. The ungulates were forced to adapt their foraging habits to seasonal growth in vegetation, Herbivores, such as primates, dependent on year-round availability of fruit, disappeared from the northern continents at this time, migrating with the retreating tropical forests to the southern continents. As there was no direct land route between northern and southern continents at this time, larger and less agile animals such as ungulates apparently failed to do so. In the absence of a year-round supply of non-fibrous growing parts of the plants a considerable increase in body size occurred amongst the artiodactyls, with one lineage (the suborder Suina) remaining as omnivores and the others (ruminants) becoming specialized herbivores. With this initial increase in body size, it was possible for the ruminant artiodactyls to evolve their characteristic foregut site of fermentation to accompany this more fibrous diet as the disadvantages of being a ruminant (ie inability to avoid fermenting all food with a consequent loss of nutrients) was outweighed by the greater capacity to take in food and lower energy demands per unit of volume of the larger animal. A rumen may have been additionally advantageous in enabling them to conserve protein by recycling urea, and also to detoxify plant secondary compounds.

The evolutionary radiation of the ruminant artiodactyls had little effect on the equids, who had always taken a more fibrous diet than other perissodactyls, and who now continued to specialize on diets too coarse for ruminants to handle. However, the tapir radiation was badly affected. Although the tapirs managed to survive the Tertiary, their real adaptive response to the

▼ **Oligocene ungulates.** The Oligocene (38–26 million years ago) was a period of major change in the world's climate and fauna. The climate became cooler, with an ice cap forming at the South Pole and lowering the sea levels all over the world. The dense lowland forests gave way to more open woodland. In this changing environment many archaic mammals became extinct while the ancestors of modern forms made their first appearance. The predominance of odd-toed ungulates began to wane, with the even-toed forms developing in size and diversity. Nevertheless, some of the odd-toed ungulates also grew in size, the horses, the brontotheres and the giant hornless rhinoceroses of Asia, such as *Indricotherium*, being typical. Among the even-toed forms were the first ruminants, creatures such as the small oreodont *Merycoidodon*. (**1**) *Indricotherium*, a giant hornless rhinoceros. (**2**) *Mesohippus*, a three-toed horse. (**3**) *Cainotherium*, a hare-like small ruminant. (**4**) *Hyracodon*, a three-toed rhino. (**5**) *Merycoidodon*, an oreodont. (**6**) *Poebrotherium*, an early camelid. (**7**) *Brontops*, a giant titanothere.

environmental changes of the late Eocene was to give rise to the rhinos, whose larger body size, and hence lower food requirements per unit body weight, rendered them immune from direct ruminant competition, and made them better able to cope with seasonally variable vegetation.

The subtropical woodlands of the Northern Hemisphere during the Oligocene (38–26 million years ago) were good habitat for the omnivorous and larger browsing ungulates. This time represented the peak of rhino diversity in the Northern Hemisphere. Besides the "true" rhino lineage that survives today, there were large hippo-like amynodont rhinos, small pony-sized hyracodont rhinos, and small tapir-like true rhinos called diceratheres, none of which survived past the early Miocene. Omnivorous suoid entelodonts and anthracotheres were common across the Northern Hemisphere, and in North America hyrax-like tylopod oreodonts were the dominant ungulates.

In contrast to the Oligocene radiation of rhinos, suoids and oreodonts, the equids and ruminant artiodactyls were all of small body size and limited diversity. In North America, the equids were all three-toed browsers with low-crowned teeth, although there was a moderate diversity of small camelids that already showed the long legs and neck typical of open habitat animals. In addition, small traguloids (chevrotains) were abundant, and horns were developed in the camel-related protoceratids.

The early Miocene (25 million years ago) saw a gradual climatic change in the northern latitudes, with the subtropical woodland replaced by a more open savanna type of habitat. The entelodont suoids, and the amynodont and hyracodont rhinos became extinct, the oreodonts greatly reduced in numbers and diversity, and the anthracotheres confined to the tropical regions of Africa and southern Asia. The North American continent suffered early extinction of rhinos and chalicotheres during the Miocene, probably because it was still isolated from South America, and provided no tropical refuge for animals during periods of climatic fluctuations. However, Eurasia was in contact with Africa by this time, many Eurasian ungulate families made their first appearance in Africa, and the Old World suffered less dramatic reduction in the numbers of large browsing ungulates during the Tertiary.

The Miocene climatic changes encouraged the spread of the grasslands, and with them the explosive radiation of the equids and ruminant artiodactyls. In the Old World the horned pecoran families (ancestors of giraffes, deer, musk deer, bovids), larger than the earlier hornless forms, appeared and diversified. In North America, the camelids also increased in size and

▲ **Miocene ungulates.** All the modern groups of hoofed mammals evolved during the Miocene (26–7 million years ago) and the following epoch, with even-toed ungulates now outnumbering odd-toed. (**1**) *Moropus*, a claw-bearing chalicothere. (**2**) *Synthetoceras*, a grotesque four-horned protoceratid ruminant. (**3**) *Alticamelus*, a giraffe-like camelid. (**4**) *Cranioceras*, a deer-like protoceratid ruminant. (**5**) *Hypohippus*, a woodland browsing horse. (**6**) *Teloceras*, a hippo-like amphibious rhinoceros. (**7**) *Blastomeryx*, a primitive deer-like ruminant. (**8**) *Dinohyus*, a giant pig-like entelodont. (**9**) *Dicrocerus*, a primitive cervoid or deer.

▶ **Pliocene ungulates.** The Pliocene (7–2 million years ago) saw the emergence of the first great grass plains, dominated by large herds of antelopes, mastodonts (primitive ungulates) and three-toed horses. The even-toed ungulates continued their dramatic development, especially of deer, giraffids and antelopes, included in which were a large number of primitive bovids. (**1**) *Paleotragus*, an okapi-like giraffid. (**2**) *Helladotherium*, a short-necked giraffid. (**3**) *Paleoreas*, an early antelope. (**4**) *Dicerorhinus*, a small long-legged rhinoceros. (**5**) *Hipparion*, a three-toed horse. (**6**) *Pliohippus*, the first one-toed horse. (**7**) *Alticornis*, a typical early pronghorn. (**8**) Duiker, a small antelope.

diversified into a number of lineages, including forms paralleling African ungulates like gazelles, giraffes and eland, and the newly immigrant deer-like artiodactyls (cervoids) diversified into forms paralleling the smaller grazing and browsing African antelopes. While bovids are the most diverse and geographically abundant of the pecoran families today, in the global woodland savanna habitat of the Miocene the cervoids predominated. Pigs and giraffes were also more diverse during the Miocene than they are today. The giraffes, which first appeared in the early Miocene of Africa, initially consisted of two distinct lineages: the long-necked, high-browsing giraffes, surviving in reduced diversity today and especially—in the Pliocene of southern Eurasia—a lineage of low-level browsing giraffes, the sivatheres, with shorter necks, higher-crowned teeth, and moose-like palmate horns which became extinct about a million years ago.

The first hypsodont grazing equids appeared at the close of the early Miocene in North America, and the late Miocene represented the peak of equid diversity, when there were at least six genera sharing the same habitats in North America, including a large persistently browsing form (*Hypohippus*) and a gazelle-sized three-toed grazing form (*Calippus*). It seems likely that the North American savannas were more arid than the African ones, resulting in a type of tough vegetation more suitable for sustaining equids than ruminant artiodactyls. In contrast to the Old World ruminants, none of the medium- to large-sized North American camelids or cervoids were grazers, and the Miocene radiation of the grazing equids in North America can be seen as equivalent to the more recent radiation of grazing bovids in the African savannas. The true Old World savanna fauna began to emerge in the late Miocene with the invasion of the three-toed grazing equid *Hipparion* and diversification of the large grazing bovids.

The diversity of the endemic North American ungulates declined in the Plio-Pleistocene, with the savanna habitat giving way to prairie. However, species of the larger camelids and one-toed grazing equids retained a moderate diversity and abundance, both families invading South America when these continents became connected in the early Pleistocene. About 10,000 years ago an unknown catastrophe wiped out the endemic North American ungulates, leaving the pronghorn as the only survivor of the late Tertiary radiations, and leaving us with a reduced impression of the suc-

cess of equids and camelids in diversifying.

In contrast, the Old World retained a diversity of tropical and subtropical habitats throughout the Plio-Pleistocene, and it was during this time that the bovids experienced their explosive radiation. These subtropical savanna faunas, containing not only bovids but a diversity of giraffids and the equid *Hipparion*, were widespread in Africa and Southern Eurasia during the Pliocene, but in the Pleistocene Ice Ages the geographical range and diversity of the giraffids decreased, whereas those of the more temperate habitat cervids (true deer) expanded with giant forms such as the Irish "elk" *Megaceros* and the giant moose *Cervalces*. Cervids and bovids migrated into North America in the Pleistocene, but cervids alone reached South America. Both equids and rhinos were abundant as temperate and cold-adapted forms in the Pleistocene of Eurasia and included the massive Woolly rhino *Coelodonta* and the rhino *Elasmotherium* which had a long single horn and continuously growing cheek-teeth. However, the ranges of Old World perissodactyls were greatly reduced, possibly due to the influence of emerging human populations.

Ecology and Behavior

The social organization of different ungulate species is a consequence of the body size of the animal, its diet, and the structure of the habitat in which it lives. Such factors determine the relative advantages and disadvantages of group feeding, taking into account, for example, predator avoidance and food availability.

Small species are vulnerable to many more predators than are large ungulates, and often cannot avoid predation by flight or self-defense. They seek to avoid detection, and most are cryptically colored. They eat buds and berries which are scattered and scarce items of high-protein content and best sought alone, preferably over familiar ground. A herd of such animals would be too scattered in its foraging to result in any advantage for predator detection or avoidance, and many such animals all feeding in one area would soon deplete all the available resources. Many small artiodactyls thus forage singly or in pairs throughout a so-called resource-defended territory. They live in tropical forest or woodland where such food is readily available, and cover is ample to hide from predators. In perissodactyls, a similar social system is seen in tapirs and the browsing rhinos (see pp56–57). For all animals of this type, contacts and conflicts between members of the same species are few, they tend to be solitary or monogamous in their reproductive behavior and show little difference in appearance or body size between the sexes.

Larger species of perissodactyls and artiodactyls can tolerate food of a lower protein content and are better able to avoid predators. Consequently they can forage in open habitats, where their food is more abundant than is that of a small forest browser. Open-country ungulates can thus benefit from membership of a herd, where cohesion and communication between a group of animals can aid in early predator detection, and where any one individual is less likely to be the victim of an attack, and, because of the greater availability of food, they are less likely to interfere with each other's feeding behavior.

▲ **Pleistocene ungulates.** The Pleistocene, which began about 2 million years ago, was the period of the ice ages in the Northern Hemisphere. There was a tendency towards gigantism in many mammalian forms. (**1**) *Coelodonta*, the Woolly rhinoceros. (**2**) *Elasmotherium*, a six-foot-horned giant rhinoceros. (**3**) *Sivatherium*, a great antlered giraffid. (**4**) *Eucladoceros*, a large "bush-antlered" deer. (**5**) *Megaceros*, the giant Irish deer. (**6**) Aurochs, the ancestral bovid of modern European cattle. (**7**) *Capra ibex*, an early form of the modern ibex.

▼ **Prehistoric frieze.** Cave paintings at Lascaux, France, dating from about 18,000BC.

Nearly all large ruminant artiodactyls are found in herds, as are the grazing perissodactyls. The exceptions among the artiodactyls are specialist forest browsers, such as the okapi and the moose (see pp92–93), which seek refuges—water, dense forest—to escape predators.

Among the larger artiodactyls, two distinct types of herds are apparent: fixed or temporary membership. In the former, closed-membership herd, the members are usually closely related, and show group defense against predators. Within this type of mixed sex herd, the males establish a dominance hierarchy to determine mating rights for the females when they come into heat. Although the females in such herds usually resemble the males in having horns, male dominance depends on size, and in such species the males are usually much larger than the females, and may continue to grow for most of their lives. Such species are mainly large-bodied open-habitat bovids, such as bison (see pp114–115), buffalo and Musk oxen.

In artiodactyl species with open-membership herds, there are some species in which mating rites are tied to territorial possession, while in others it is tied strictly to rank. These territories are often mating areas that can support at most only one male, rather than a resource area for a harem. Thus females will enter and leave the male's territory, and the most successful males are those who hold the most attractive "property." Males must challenge other males for the rights to a territory, and rarely do they hold it for long, as its defense is exhausting. Such species tend to have the most marked difference in appearance and body size between the sexes, and the males have elaborate horns or antlers to engage in continual territorial combat. Most medium- to large-sized deer and antelope belong to this category. Some species, such as the migratory wildebeest, have this type of social organization, yet appear to lack clearly marked territories which are fixed in locality. In others, such as the kob, a "lek" system is seen, in which receptive females visit the holders of conventionalized, highly contested, close-packed mini-territories for mating.

In contrast to the ruminant artiodactyls, the typical perissodactyl social system (as exemplified by equids) is one where a single male consorts with a group of females, constituting a fixed membership harem with strong interpersonal bonds. Fights between males occur more rarely than in artiodactyls, as the male horse neither defends a fixed territory, nor does he have to continually contend with other adult males within the herd. Males without harems (usually younger males) form roving bachelor herds of more variable composition. A modified version of this social system is seen in the Wild ass, Grevy's zebra and the White rhino (see pp56–57), where the males additionally defend a territorial area. However, no perissodactyl exhibits the pronounced sexual differences typical of ruminant artiodactyls.

Exclusive mating territories maintained by males are typical of medium-sized ruminants of tropical and subtropical woodland and savanna, but not of comparable perissodactyls in similar habitats. Probably a ruminant needs a smaller area than would a similar perissodactyl, because of its ability to survive on a smaller quantity of food per day. However, the situation is reversed in low productivity open-grassland when a perissodactyl would be able to maintain a smaller home range by dint of eating everything available, whereas the ruminant would have to forage further afield in order to locate food of suitable quality. Thus in the arid open grasslands, equids and the White rhino are territorial, whereas bison and buffalo maintain a non-territorial dominance hierarchy system.

Most ungulate females mature relatively quickly for their size, have long gestation periods, and bear one, large, well-developed juvenile at a birth. Pigs are an exception, in producing several piglets in a litter. All artiodactyl juveniles can see, hear, call out,

and stand to suckle soon after being born. All females, except pigs, make no nest but usually seek seclusion shortly before giving birth. Where cover is available many ungulates hide the young, the mother leaving it while she feeds. The calf hides immobile until her return. In contrast some of the largest, open-country, herd-forming species, for example wildebeest (artiodactyl) and horses (perissodactyl), which have evolved the ability to get to their feet and start moving remarkably soon after birth; some can run at near adult speed within half an hour of birth.

Despite relatively slow reproduction, artiodactyls are not very long-lived for their size; a 10-year-old impala, or a 15-year-old Red deer, would be old animals. Perissodactyls are longer-lived; up to 35 years in equids and 45 years in rhinos. As the most abundant large mammals, ungulates form the staple diet of most of the great terrestrial carnivores. For those which do not fall to predators, starvation is an annual threat since populations of many of them, as dominant herbivores in their communities, live close to the minimum yearly carrying capacity of the plant community. Although they have evolved a considerable capacity for storing energy in their body tissues, to be used when food is low, extra burdens of energy expenditure, such as suckling a calf or defending a territory, will weaken some individuals to the point where they are vulnerable to disease, predators or final starvation. At the last, starvation is inevitable when their abrasive diet finally reduces their grinding molars to eroded stumps which are useless for feeding.

Ungulates and Man

The genus *Homo* emerged near the peak of the bovid radiation and has shared the Pleistocene and recent fortunes of that family. Humans are partly responsible for the dwindling numbers of ungulate species from the Pleistocene. The dominant position of ungulates in most communities of large herbivores, their use of habitats, size, social organization and ecology, even their antlers and horns, have all exposed them to damaging interactions with humans. Yet those same characteristics have been the salvation of some.

The human ancestors whose remains have been found in eastern Africa lived in a community of ungulates similar to that of the same area today, but containing a number of now-extinct "giant" forms. Such large ungulates were at minimal individual risk from most predators. However,

although neither strong nor fast, humans excelled in three traits: perceptive and inventive intelligence; coordinated and mobile group hunting; and the use of artificial weapons, especially projectiles. These made humans the most generalized large predators in the community, able to hunt all prey in a wide variety of circumstances. From the glacial and interglacial epochs in Eurasia comes a wealth of evidence from societies culturally dependent upon the hunting of ungulates. Not only are there bones of the animals they killed, but the hunters also left their own record in the form of art: carvings, clay models and, most dramatic of all, vast cave-murals depicting many of the animals.

Small wonder, then, that the hunting of ungulates was the symbol of early heroism, and sport and trophy hunting are still widely accepted as justifiable ways of

▲ **A panorama of pastoralism.** Somali nomads with their docile and well-ordered cattle, sheep and camels at a water-hole in northeastern Kenya epitomize man's exploitation of hoofed mammals. These vast herds of domesticated stock have largely displaced wild hoofed mammals and profoundly affected the environment across much of North Africa, the Near East, Mediterranean Europe, and northern Asia.

▶ **Boar hunting.** A Persian bas-relief showing the Emperor Khusraun II (AD591–628) killing a Wild boar.

at other centers in its wide range in Asia at other dates. Cattle were domesticated by 6500BC from wild cattle or aurochs in Europe and the Near East, although other centers of domestication from rather different stock may have led to the cattle of India and East Asia. These major domestications occurred at about the same time as domestication of wheat, barley and the dog, and preceded by 2,000–4,000 years the domestication of donkeys, horses, elephants and camels. In South America, domestic breeds of llamas appeared between 4000 and 2500BC, some 2,000 years after the domestication of maize. The temporarily settled conditions of a primitive, crop-growing society would have been ideal for the domestication of captured ungulates; indeed settlement rather than nomadic hunting would have made domestication necessary to ensure meat.

Horses and donkeys were the last of the common livestock animals to be domesticated, and they have been the least affected by human manipulation and artificial selection.

The first domestic horses appeared at about 4000BC, when they may have initially been used for food, but it was not until about 2000BC that the widespread use of the horse as a means of transport came about, which caused a revolution in the human mobility race and the development of modern techniques of warfare.

Some domesticated stock remained confined to the areas in which they were domesticated (yak, or dromedaries for example). But cattle, sheep and goats, with horses, donkeys and camels, supported human groups which formed distinct, nomadic pastoral cultures, depending almost entirely on their animals, not on crops. These people lived typically on milk from their stock, supplemented occasionally by meat or rarely by blood taken without killing the beast. They spread throughout the savanna, steppe and semidesert lands of Eurasia and Africa. Their cumulative effect has been to alter grossly the environment of Mediterranean Europe, the Near East, northern Asia, and much of Africa, by the combined grazing and browsing pressure of this spectrum of stock.

These changes to the landscape affected not only humans dependent on the stock, but also the wild herbivores of the area. Pastoralism everywhere has diminished wild ungulates to the point where communities remain only where physical barriers or the risk of disease have kept out humans and their stock.

enhancing a person's status and prestige.

The crucial event in the relationship between ungulates and man was domestication—some 15 species of ungulates have been domesticated. This has certainly occurred independently with different species—domestication of South American camelids is quite separate from any Old World domestication, for example. The features which occur most commonly amongst domesticated ungulates are a tendency, for the females at least, to occur in closed-membership herds, and for males to be non-territorial. These characteristics lend themselves to herding and tending by humans. In the Old World, sheep and goats were domesticated by 7500BC (perhaps much earlier), from mouflon and Wild goat respectively. The pig was domesticated from Wild boar by 7000BC in the Middle East, but was probably domesticated independently

CJ/PJJ

ODD-TOED UNGULATES

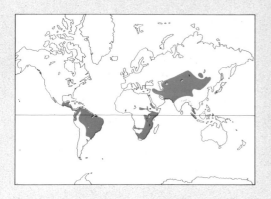

Order: Perissodactyla
Sixteen species in 6 genera and 3 families.
Distribution: Africa, Asia, S and C America.

Habitat: diverse, ranging from desert and grassland to tropical forest.

Size: head-body length from 71in (180cm) in the Mountain tapir to 145–160in (370–400cm) in the White rhino. Weight from 495lb (225kg) in the Brazilian tapir to 5,070lb (2,300kg) in the White rhino.

Horses, Asses and Zebras (family Equidae)
Seven species of the genus *Equus*: **African ass** (*E. africanus*), **Asiatic ass** (*E. hemionus*), **Domestic horse** (*E. caballus*), **Grevy's zebra** (*E. grevyi*), **Mountain zebra** (*E. zebra*), **Plains zebra** (*E. burchelli*), **Przewalski's horse** (*E. przewalskii*).

Tapirs (family Tapiridae)
Four species of the genus *Tapirus*: **Baird's tapir** (*T. bairdi*), **Brazilian tapir** (*T. terrestris*), **Malayan tapir** (*T. indicus*), **Mountain tapir** (*T. pinchaque*).

Rhinos (family Rhinocerotidae)
Five species in 4 genera: **Black rhinoceros** (*Diceros bicornis*), **Indian rhinoceros** (*Rhinoceros unicornis*), **Javan rhinoceros** (*Rhinoceros sondaicus*), **Sumatran rhinoceros** (*Dicerorhinus sumatrensis*), **White rhinoceros** (*Ceratotherium simum*).

T HE odd-toed ungulates, the Perissodactyla, are a small order of mammals today, with the horses, and the closely related asses and zebras, being the only well-known and widely spread members of the group. The familiarity of the horse is, of course, largely a consequence of its domestication and use by humans, first as a means of transport, warfare and agricultural labor, and today predominantly for recreation and sport. Populations of wild equids are limited in their abundance and geographical distribution.

The other members of the order are animals that one does not usually associate with the graceful open-country horse: the ponderous and endangered rhinos of Africa and Asia, and the rare and elusive tapirs of the tropical forest of Malaysia and South America. But details of the anatomy of these animals, and overall similarities in their behavior and physiology, can be shown to unite them in a single order.

Today perissodactyls apparently run a poor second to the even-toed ungulates, artiodactyls, in terms of numbers of species, geographical distribution, variety of form and ecological diversity. But our present-day viewpoint belies the success of the odd-toed ungulates over geological time. They were the dominant ungulate order during the early Tertiary (54–25 million years ago), and their subsequent decline is more likely to have been due to climatic factors than to direct competition with artiodactyls.

Perissodactyls first appeared in the late Paleocene (58 million years ago) in North America, and by the early Eocene (54 million years ago) five out of the six known families were evident. Living perissodactyl families are the Tapiridae (tapirs), the Rhinocerotidae (rhinos) and the Equidae (horses, asses and zebras). Tapirs and rhinos are closely related, with rhinos representing an offshoot of the tapir family in the late

Eocene. These two families are frequently grouped together as the suborder Ceratomorpha, distinguishing them from the equids in the suborder Hippomorpha. Rhinos are characteristically heavy bodied and show adaptations of the limbs for weight bearing (graviportal). In contrast, equids showed modifications for running from the start of their evolution, with a progressive tendency to lengthen the limbs and reduce the lateral digits.

Living tapirs are medium-sized animals, principally inhabiting tropical and subtropical woodland in Malaysia, and Central and South America, where they feed on a mixture of browse material and fruit. Tapirs have only inhabited South America since the Pleistocene (2 million years ago), and until the Pleistocene Ice Ages were also found in Europe and North America. Living rhinos are all large, although pony- and tapir-sized rhinos were common in the early Tertiary. They are found today in Africa and Southeast Asia, and occupy a variety of habitats ranging from forest to grassland in tropical and subtropical areas.

Their diet varies from browse, through a mixture of grass and browse, to grass exclusively, depending on the species. Rhinos first appeared in Africa at the start of the Miocene (25 million years ago), were found in North America until the end of the Miocene (7 million years ago), and were common throughout Eurasia until the end of the Pleistocene (10,000 years ago).

Living equids are medium sized, and they are all specialist grazers, inhabiting grassland ranging from woodland savanna to arid prairie in temperate and tropical regions. They were widely distributed in all continents except Australasia until the end of the Pleistocene, but today are found only in Africa and parts of Asia, although feral populations of horses and asses flourish in Europe, Australia, western North America

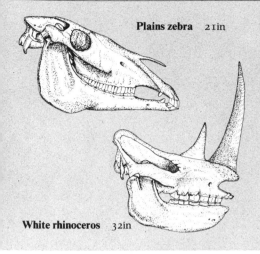

Plains zebra 21in

White rhinoceros 32in

Skulls of Odd-toed Ungulates

Odd-toed ungulates may have low- or high-crowned cheek-teeth, but they all tend to have molarized premolars, with the molar cusps coalescing to form cutting ridges. Tapirs retain the complete mammalian dental

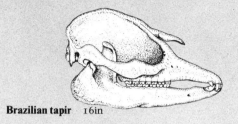

Brazilian tapir 16in

formula, and have a fairly generalized skull shape. Rhinos are characterized by an especially deep and long occipital region (rear of the skull), associated with the large mass of neck musculature needed to hold up their massive head, and by their characteristically projecting and wrinkled nasal bones, on which the keratinous horns are attached. The incisors in rhinos are reduced to two upper and one lower, and tend to be lost entirely in the grazing species. The skulls of equids like the Plains zebra show a great elongation of the face, the posterior position of the orbits (to allow room for the exceedingly high-crowned cheek-teeth), the complete post-orbital bar, and the exceedingly deep and massive lower jaw.

▼ Division of the spoils among late Eocene perissodactyls in North America. The surviving perissodactyl species are rarely found in the same habitats but 40 million years ago they, and species that are now extinct, were the dominant hoofed mammals in North America. TOP LEFT The extinct brontotheres took high-level browse, including about 50 per cent fruit. TOP RIGHT The extinct chalicotheres took high-level browse of moderately good quality, plus some fruit. MIDDLE Tapirs took low-medium-level browse, of good quality, preferably green shoots. BOTTOM RIGHT Equids took moderately high-fiber low-level browse, supplementing protein intake with buds, berries, some fruit etc. BOTTOM LEFT Rhinos took moderately high-fiber browse.

and South America. The center of equid evolution was North America, although an offshoot to the subsequently successful North American lineage was present in Eurasia in the Eocene. True equids (genus *Equus*) did not apear in the Old World until the Pleistocene (2 million years ago), but previous invasions of Eurasia were made by a three-toed browsing horse in the early Miocene, and of Eurasia and Africa by a three-toed grazing horse in the late Miocene, which survived into the Pleistocene and was found for a time alongside true equids. Equids became extinct in North America only about 10,000 years ago.

In addition to the three living families of perissodactyls, three of the original families are now extinct. Grouped with the equids in the suborder Hippomorpha are the rhino-like brontotheres, which lived in North America and Asia in the Eocene and early Oligocene, and tapir-like paleotheres, which lived during the same period in Europe. The family Chalicotheriidae occupied its own suborder Ancylopodia, as chalicotheres were sufficiently different from other peris-sodactyls to warrant separate classification. They were largish animals, ranging from the size of a horse to the size of a giraffe, and had secondarily substituted claws for hooves on their feet. Chalicotheres were initially found in both North America and Eurasia; they first appeared in Africa along with rhinos in the Miocene, and became extinct in North America in the middle Miocene, but persisted into the Pleistocene in Asia and Africa. However, speculation abounds about their possible survival into more recent times. Chalicothere-like animals, with a horse-like head and bear-like feet, appear on plaques in Siberian tombs dating from the 5th century BC, and it has been suggested that the mysterious Kenyan "Nandi bear" could be a surviving chalicothere. CJ

HORSES, ASSES AND ZEBRAS

Family: Equidae
Seven species in 1 genus.
Order: Perissodactyla.
Distribution: E Africa, Near East to Mongolia.

Habitat: from lush grasslands and savanna to sandy and stony deserts.

Size: ranges from head-body length 79in (200cm), tail length 16.5in (42cm) and weight 605lb (275kg) in the African ass to head-body length 108in (275cm), tail length 19in (49cm), and weight 885lb (405kg) in the Grevy's zebra. Gestation: about $11\frac{1}{2}$ months, but $12\frac{1}{2}$ in Grevy's zebra.

► **Fighting Plains zebras** at Amboseli National Park, Kenya. Such contests between males are common when females are ready to mate.

▼ **Plains zebras drinking** at a water-hole at Umfolozi National Park, South Africa.

Ever since horses were first domesticated in Asia, five thousand years ago, they have served mankind as a beast of burden and as a means of transport; they have helped him till the soil and wage war; they have provided him with recreation and even companionship; and with mane waving and hooves thundering, they have served as a symbol of power, grace and freedom. But the domesticated horse is just one member of the once diverse family Equidae. Only six other species survive, some precariously.

All equids are medium-sized herbivores with long heads and necks and slender legs that bear the body's weight only on the middle digit of each hoofed foot. They possess both upper and lower incisors that clip vegetation and a battery of high-crowned, ridged, cheek-teeth that are used for grinding. The ears are moderately long and erect, but can be moved to localize sounds and to send visual signals. A mane covers the neck, but only in the Domestic horse does it fall to the side. On the other species it stands erect. All equids have long tails, which are covered with long flowing hair in horses, and short hair only at the tip of asses and zebras.

The species differ somewhat in size, and males are generally 10 percent larger than females. But the most striking feature that distinguishes species is coat color, the zebra's stripes being the most dramatic example (see pp46–47). The coats of horses and asses are more uniform in color: dun in Przewalski's horse, from tan to gray in asses.

Equids' eyes are set far back in the skull, giving a wide field of view. Their only blind spot lies directly behind the head, and they even have binocular vision in front. They probably can see color and although their daylight vision is most acute their night vision ranks with that of dogs and owls. They can detect subtle differences in food quality. Males use the flehmen or lip-curl response to assess the sexual state of females, and the vomeronasal or Jacobson's organ which is used in this is well developed. Equids can also detect sounds at great distance and by rotating their ears can locate the source without changing body orientation.

Moods are often indicated visually by changes in ear, mouth and tail positions. Smell assists individuals in keeping track of the movements of neighbors, since urine and feces bear social odors, but social contact is effected primarily by sounds. In horses and Plains zebra, mothers whinney when separated from their foals and nicker to warn them of danger. Males often nicker to declare their interest in a female and they squeal to warn competitors that further escalated combat is imminent. In asses and Grevy's zebra, males often bray when fighting, or calling to each other over long distances.

The Plains zebra occupies the lushest environments—the grasslands and drier savannas of East Africa from Kenya to the Cape. As its name implies, the Mountain zebra is restricted to two mountainous regions of southwest Africa where vegetation is abundant. The remaining species live in more arid environments with sparsely distributed vegetation. Przewalski's horse inhabits the semi-arid deserts of Mongolia, while the Asian wild ass inhabits the most arid deserts of central Asia and the Near East. The African wild ass, the least horse-like of the equids, roams the rocky deserts of North Africa. Generally, the ranges of the species do not overlap, the only exception being Grevy's and Plains zebra, which coexist in the semi-dry thorny scrubland of northern Kenya. Only the Domestic horse is found worldwide, and it has spawned feral populations in North America on the western plains and on east coast barrier islands, and in the mountains of western Australia.

The earliest of the horse-like ancestors, *Hyracotherium*, appeared in the Eocene, about 54 million years ago; it was a small dog-sized mammal which browsed on low shrubs of the forest floor, and had low-crowned teeth without the complex enamel ridges of modern equids. It had already lost two hind toes on its hindfeet and one on its forefeet, but the feet were still covered with soft pads. When grasses appeared in the Miocene, equids began to radiate. Continuously growing teeth with high crowns, complex grinding ridges and cement-filled interstices evolved with the opening out of the habitat, and the need to run from predators and to travel long distances in search of food and water led to many changes in body shape. Overall body size increased, which reduced relative nutritional demands. By the early Pleistocene (2 million years ago), the one-toed equids had spawned the genus *Equus*, which rapidly spread all over the world.

As environments changed, populations became isolated, giving rise to most of the living species. The first to split off from the equid stem after becoming single-toed was the Grevy's zebra, which, despite its stripes, is only distantly related to the other two zebra species. The only species that probably did not originate via geographic isolation is

the ancestor of the Domestic horse. It is thought to be directly descended from some mutant Przewalski's horses.

All equids are perissodactyls that forage primarily on fibrous foods. Although horses and zebras feed primarily on grasses and sedges, they will consume bark, leaves, buds, fruits and roots, which are common fare for the asses. Equids employ a hindgut fermentation system, in which plant cell walls are only incompletely digested but processing is rapid. As long as they are able to ingest large quantities of food, they can achieve extraction rates equal to those of ruminants. Because forage quality does not affect the process, equids can sustain themselves in more marginal habitats and on diets of lower quality than can ruminants. Equids do spend most of the day and night foraging. Even when vegetation is growing rapidly, equids forage for about 60 percent of the day (80 percent when conditions worsen). Although equids can survive on low-quality diets, they prefer high-quality, low-fiber food.

Equids are highly social mammals that exhibit two basic patterns of social organization. In one, typified by the two horse species, as well as the Plains and Mountain zebra, adults live in groups of permanent membership, consisting of a male, a few females remain in the same harem throughout their adult lives. Each harem has a home range, which overlaps with those of neighbors. Home range size varies, depending on

---

Abbreviations: HBL = head-body length. TL = tail length. wt = weight.
⬚E⬚ Endangered. ⬚*⬚ CITES listed.
⬚V⬚ Vulnerable.

Subgenus *Equus*

N and S America, Mongolia, Australia. General body form variable, in part due to domestication. Long tails with hairs reaching to the middle of leg. Usually solid coat colors.

Przewalski's horse ⬚E⬚

Equus przewalskii
Przewalski's, Asiatic or Wild horse.

Mongolia near Altai mountains. Open plains and semidesert. HBL 83in; TL 35in; wt 770lb. Coat: dun on sides and back, becoming yellowish white on belly; dark brown erect mane with legs somewhat grayish on inside; thick-headed, short-legged and stocky; regarded as true wild horse.

Domestic horse

Equus caballus
Domestic or Feral horse.

N and S America and Australia. Open and mountainous temperate grasslands, occasionally semideserts. HBL 79in; TL 35in; wt 770–1,540lb. Coat: sandy to darkish brown; mane falls to side of neck. Dozens of varieties. Feral forms thick-headed and stocky, domestic breeds slender-headed and graceful-limbed.

Subgenus *Asinus*

N Africa, Near East, W and C Asia. Sparsely covered highland and lowland deserts. Horse-like, or even stockier, with long pointed ears, tufted tails, uneven mane and small feet.

African ass ⬚E⬚

Equus africanus

Sudan, Ethiopia and Somalia. Rocky desert. HBL 79in; TL 16in wt 605lb. Coat: grayish with white belly and dark stripe along back. Nubian subspecies with shoulder cross; Somali subspecies with leg bands. Smallest equid with narrowest feet. Subspecies: 3

Asiatic ass ⬚V⬚ ⬚*⬚

Equus hemionus

Syria, Iran, N India, and Tibet. Highland and lowland deserts. HBL 83in; TL 19in; wt 640lb. Coat: summer, reddish brown, becoming lighter brown in winter; belly is white and has prominent dorsal stripe. Most horse-like of asses with broad round hoofs and larger than African species. Subspecies: 4.

Subgenus *Hippotigris*

E and S Africa. Resemble striped horses.

Plains zebra

Equus burchelli
Plains or Common zebra.

E Africa. Grasslands and savanna. HBL 90in; TL 20in; wt 518lb. Coat: sleek with broad vertical black and white stripes on body, becoming horizontal on haunches. Always fat looking, but short-legged and dumpy. Subspecies: 3.

Mountain zebra ⬚V⬚

Equus zebra

SW Africa. Mountain grasslands. HBL 85in; TL 20in; wt 570in. Coat: sleeker than Plains zebra with narrower stripes and a white belly. Thinner and sleeker than Plains zebra with narrower hoofs and dewlap under neck. Subspecies: 2.

Subgenus *Dolichohippus*.

Grevy's zebra ⬚E⬚ ⬚*⬚

Equus grevyi
Grevy's or Imperial zebra.

Ethiopia, Somalia and N Kenya. Subdesert steppe and arid bushed grassland. HBL 108in; TL 19in; wt 990lb. Coat: narrow vertical black and white stripes on body, curving upwards on haunches; belly is white and mane prominent and erect. Mule-like in appearance with long narrow head and prominent broad ears.

the quality of the habitat, and varies in Plains zebra from 31–97sq mi (80–250sq km) in the Ngorongoro crater to over 135sq mi (350sq km) during the rainy season in the Serengeti. Since the habitat deteriorates during the dry season, the Serengeti zebra harems congregate and migrate *en masse* about 60mi (100km) to different habitats, where home ranges are often as large as 230sq mi (600sq km).

The second social system, typified by the asses and Grevy's zebra, involves more ephemeral adult associations, rarely lasting longer than a few months. Temporary aggregations of one or both sexes are common, but most adult males live alone within large territories. For Grevy's zebra these vary in size from 0.8–3.9sq mi (2–10sq km), but for the asses they can be as large as 5.8sq mi (15sq km). Within territorial boundaries, which are marked with large piles of dung, owners obtain exclusive mating access to receptive females that wander through them. In both systems, surplus males live together in bachelor groups.

Social systems involving temporary groupings and solitary territorial males occur in drier habitats and in areas where resources are distributed in a more patchy fashion. Small, widely scattered patches of low-quality vegetation preclude the formation of long-lasting associations by intensifying competition among females. And without female groups to defend, males must defend large areas containing the resources females require if they are to obtain a disproportionate share of the matings. Only larger, more evenly distributed resources allow females to feed in permanent groups, thus enabling harems to form.

▲ **Representative species of horses, asses and zebras.** (1) Przewalski's wild horse (*Equus przewalskii*), the ancestor of all domestic horses, showing the stallion's bite threat. (2) A female African ass (*Equus africanus*), showing the kick-threat, with its ears held back. (3) A male onager, a subspecies of the Asiatic ass (*Equus hemionus*), adding to a dung pile, as a territorial mark. (4) The kiang, the largest subspecies of Asiatic ass, showing the flehmen reaction after smelling a female's urine. (5) A young male Mountain zebra (*Equus zebra*) showing a submissive face to an adult male. Note the dewlap and the grid-iron rump pattern. (6) A female Grevy's zebra (*Equus grevyi*) in heat and showing the receptive stance, with hindlegs slightly splayed and tail raised to one side. (7) A male Plains zebra (*Equus burchelli*) driving mares in a characteristic low-head posture, with ears held back.

Even for equids that maintain long-lasting associations, their groups are different from those of most mammals. Typically, daughters remain with their mothers, creating groups composed of close kin, but in equids groups are composed of non-relatives, since both sexes leave their natal area. Females emigrate when they become sexually mature at about two years old, and neighboring harem or bachelor males attempt to steal them. Males disperse by the fourth year to form bachelor associations. Only after a number of years in such associations are males able to defend territories, steal young females, or displace established harem males.

Mares usually bear only one foal. Only in the Grevy's zebra does gestation last slightly more than one year; since females come on heat 7–10 days after bearing a foal, mating and birth occur during the same season, coinciding with renewed growth of the vegetation. Reproductive competition among males for receptive females is keen. This begins with pushing contests, or ritualized bouts of defecating and sniffing, but it is not uncommon for contests to escalate, with animals rearing and biting at necks, tearing at knees or thrusting hindlegs towards faces and chests. In contrast, amicable activities, such as mutual grooming, cement relationships among females. There are, however, dominance hierarchies among females, and high rank confers substantial benefits, including first access to water and superior vegetation.

Young are up and about within an hour of birth. Within a few weeks they begin grazing, but are generally not weaned for 8–13 months. Females can breed annually, but most miss a year because of the strains of rearing foals.

Despite the proliferation of domestic horses, their wild relatives are in a precarious situation. No Przewalski's horses have been seen in their natural habitat since 1968, and only about 200 individuals, scattered among the world's zoos, separate the species from extinction. Many feral populations of Domestic horses which roam freely are treated as vermin. As for zebras, only the Plains zebra is plentiful and occupies much of its former range. Populations of Mountain zebra are small and are protected in national parks, but those of Grevy's zebra have been drastically reduced, since their beautiful coats fetch high prices. It would be tragic if these wonderful creatures were to go the way of the South African quagga, a yellowish-brown zebra with stripes only on the head, neck and forebody, which was exterminated in the 1880s.   DR

The Zebra's Stripes

An aid to group cohesion?

The term zebra has no taxonomic meaning. It describes three equines that have stripes. On fossil and anatomical evidence, the Grevy's and Mountain zebras are as distantly related to each other as they are to horses and asses. However, the ancestor of all equines was probably striped.

What do zebras have in common apart from their stripes? They live in Africa and they are exceptionally social: groupings range from two to several hundreds; they disperse or aggregate in response to the vagaries of pasture, water and climate. Like all horses, they are nomadic grazers of coarse grasses. They are active, noisy and alert. They never attempt to conceal themselves or to "freeze" in response to predators and they prefer to rest, grouped, in exposed localities where they have the advantage of a good view at the cost of being conspicuous themselves. The widespread theory that their stripes are camouflage is therefore contradicted by the zebra's behavior.

Opinion is divided as to whether the stripes are a visual or chemical device. Proponents of the latter think that stripes may help the animals to regulate their body temperature, yet zebras live under many different climates, and the stripe patterns show no variation with climate. A suggestion that stripes may have evolved to deter harmful flies is based on the observation that signals of both chemical and optical origin influence flies (yet insects are no hazard over most of the zebras' very wide range of habitats).

The "visual" theorists try to imagine the effect of stripes on pests and predators. One ingenuous explanation has it that charging lions are unable to single out an individual because it merges with others in the herd; another suggests that the lion is dazzled or miscalculates its imaginary last leap. These theories founder on the observable confidence with which lions kill zebras and on the fact that in those places for which there are records zebras are killed broadly in proportion to their relative abundance.

An alternative approach has been to consider the intrinsic optical properties of stripes in relation to the physiology of the optical centers of the brain, rather than look for resemblances to anything else. Explanations of the stripes' function and the selective pressures that maintain them may be better sought in the behavior of those animals that are most exposed to seeing the stripes—not a passing predator, but the zebras themselves.

Research on vertebrate vision suggests that several kinds of primary nerve cells in the visual system are excited by crisp black and white stripes, notably the detectors of tonal contrasts, spatial frequencies, linear orientation, "edges" and the "flicker" effect of moving edges. So zebras within a herd cannot escape the visual stimulation of stripes, and there is evidence that they actively seek it. Zebras walk towards, stop and stand near one another with great consistency and show few signs of discrimination between individuals or sexes. They presumably register the totality that signals a particular fellow zebra, but their responses to artificially striped panels suggest that their attraction to stripes is more or less automatic.

The mechanism may originate in a transfer from relationships in infancy based mainly on touch (tactile) to looser, visually based associations in nibbling and grooming one another, but this is a one-to-one affair and tends to be restricted to mother-young sibling relationships.

As a foal grows older, its contacts become more numerous and casual. It is at this stage that a clear separation takes place between the tactile and visual components of the young zebra's behavior; it begins to make empty grooming gestures. Prompted by the approach of other zebras, this ritualized grimace is an exaggerated greeting that helps neutralize aggression. At a lower intensity, passive head-nodding and nib-

▲ **A black zebra?** The markings of zebras are unique to each individual and some startling patterns occur, as in this Chapman's zebra, a subspecies of the Plains zebra, with the upper parts almost solid black.

▶ **Reversed stripes.** The normal black-on-white pattern is reversed in this Plains zebra.

▼ **Visual dazzle.** A large herd of zebras presents a visually disturbing appearance to human eyes, rather like the effects deliberately created in op-art paintings, but these patterns seem to be restful to zebras.

bling at nothing are a common response to being surrounded by stripes.

It may be the association of visual stimulation with the security of mother and family that makes any other zebra attractive. This has an obvious utility whenever there are dense aggregations on temporary pastures, resting grounds or at water-holes.

The link with grooming is important for explaining how stripes may have evolved and why they are found in equines but are generally absent in the even-toed ungulates, where grooming is not the main social adhesive. Many animals have visual "markers" to direct companions to particular parts of the body. In horses and zebras the preferred area for grooming is the mane and withers. Extreme bending at the base of the neck causes skin wrinkles, so it is possible that the evolutionary origins of stripes lie in enhancement of this natural characteristic. Once the optical mechanism was established on this small target area, its effectiveness would have been enhanced by spreading over the entire animal. Significantly, the three contemporary species are most alike in the flat panel area of neck and shoulder. All horses are follow-my-leader travelers and have rump patterns which are characteristic for each species.

For three very different equines to maintain crisp, evenly spaced black and white stripes there must be very strong selective pressure in favor of stripes. If animals with defective patterns consistently fail to achieve positive social relationships, their overall fitness is likely to suffer, and this could be the explanation for abnormal patterns being so rare (for example only seven black, white, dappled or marbled zebras out of one sample of several thousand animals, skins and photographs). When a black zebra was seen, in the Rukwa Valley, Tanzania, it tended towards a peripheral position whenever its group was buzzed by an aircraft or approached by a vehicle. It is possible that "visual grooming" masks the latent animosities of zebras, and a resurgence of intolerance may be a part of the selective forces that eliminate unstriped zebras.

But what of the zebras that have lost their stripes? Crisp black and white stripes can only be maintained in sleek tropical coats. Only high densities in relatively rich habitats would warrant such a conspicuous socializing mechanism. Stripes become redundant in low-density desert-dwelling asses and they become inoperable in very cold climates with annual molts into shaggy winter coats; hence the breakdown of striping in horses and the Cape quagga. JK

TAPIRS

Family: Tapiridae
Four species of the genus *Tapirus*.
Order: Perissodactyla.
Distribution: S and C America; SE Asia.

Size: head-body length
71–98in (180–250cm); tail
2–4in (5–10cm); shoulder
height 29–47in (75–120cm);
weight 500–660lb
(225–300kg).

Brazilian tapir ☀

Tapirus terrestris
Brazilian or South American tapir.
Distribution: S America from Colombia and
Venezuela south to Paraguay and Brazil.
Habitat: wooded or grassy with permanent
water supply.
Coat: dark brown to reddish above, paler
below, short and bristly; low narrow mane.
Gestation: 390–400 days.
Longevity: 30 years.

Mountain tapir ⊻

Tapirus pinchaque
Mountain Woolly, or Andean tapir.
Distribution: Andes mountains in Colombia,
Equador, Peru and possibly W Venezuela up to
altitudes of 14,750ft (4,500m).
Habitat: mountain forests to above treeline.
Coat: reddish brown and thicker than other
species; white chin and ear fringes.
Gestation and longevity: as for Brazilian tapir.

Baird's tapir ⊻

Tapirus bairdi
Distribution: Mexico through C America to
Colombia and Equador west of Andes.
Habitat: swampy or hill forests.
Gestation and longevity: as for Brazilian tapir.
Coat: reddish brown and sparse; short thick
mane, white ear fringes.

Malayan tapir ⊟

Tapirus indicus
Malayan or Asian tapir.
Distribution: Burma and Thailand south to
Malaya and Sumatra.
Habitat: dense primary rain forests.
Size: head-body length 88–98in (220–250cm);
tail 2–4in (5–10cm); shoulder height
34–42in (90–105cm); weight 550–660lb
(250–300kg). Coat: middle part of body white,
fore and hind parts black.
Gestation: 390–395 days.
Longevity: 30 years.

⊻ Vulnerable. ⊟ Endangered. ☀ CITES listed.

Tapirs are among the most primitive large mammals in the world. There were members of the modern genus *Tapirus* roaming the Northern Hemisphere 20 million years ago and their descendants have changed little. Their scattered relict distribution is often cited as evidence for the existence of the supercontinent of Gondwanaland, on the assumption that tapirs reached their present homes overland before the continents drifted apart.

Their curious appearance has led people to liken tapirs to pigs and elephants, but in fact their closest relatives are the horses and the rhinoceros. All tapirs have a stout body, slightly higher at the rump than the shoulder, and the limbs are short and sturdy. This compact streamlined shape is ideal for pushing through the dense undergrowth of the forest floor, and similar body forms have evolved independently in two other quite unrelated South American species, the peccary (see pp64–65) and capybara, which share the same habitat.

The neck of tapirs is short and the head extends into a short, fleshy trunk derived from the nose and upper lip, with the nostrils at the tip. This small proboscis helps them to sniff their way through the jungle and is a sensitive "finger" used to pull leaves and shoots within reach of the mouth. The ears protrude and are often tipped with white, and their hearing is good, although not as acute as their sense of smell. Vision is less important for these nocturnal animals and the eyes are small and lie deep in the socket, well protected from thorns. Baird's tapir and the Brazilian tapir both have short, bristly manes extending along the back of the neck, protecting the most vulnerable part of the body from the deadly bites of the main predator, the jaguar. The dramatic coloration of the Malayan tapir—middle part of the body white, fore and hind parts black—is also a protection against predators: the patches break up and obscure the body outline so that nocturnal predators fail to detect their prey. Tapir skin is tough and covered with sparse hairs; only the Mountain tapir has a thick coat, which protects it from the cold.

Tapir tracks, often the only evidence one sees of the animal's presence, are characteristically three-toed, although the forefeet have four toes and the hindfeet three. The fourth toe, which is slightly smaller than the others, and placed to one side, higher up on the foot, is functional only on soft ground. All toes have hooves and there is also a callous pad on the foot which supports some of the weight.

Tapirs are forest dwellers, active mainly at night when they roam into forest clearings and along river banks to feed. They are both browsers and grazers, feeding on grasses, aquatic vegetation, leaves, buds, soft twigs and fruits of low-growing shrubs, but they prefer to browse on green shoots. Animals follow a zigzag course while feeding, moving continuously and taking only a few leaves from any one plant. In Mexico and South America, tapirs sometimes cause damage to young maize and other grain crops and in Malaya they are reputed to raid young rubber plantations.

Apart from mothers with young, tapirs are usually solitary. They range over wide areas. They are excellent swimmers and spend much time in water, feeding, cooling off, or ridding themselves of skin parasites. If alarmed, they may seek refuge in water and can stay submerged for several minutes; the Malayan tapir is reputed to walk on the river bottom like a hippopotamus. When the animals bathe, there is increased activity in the digestive tract and, like the hippo, they

▲ **The species of tapirs.** (1) Mountain tapir
(*Tapirus pinchaque*). (2) Brazilian tapir (*Tapirus
terrestris*). (3) Malayan tapir (*Tapirus indicus*).
(4) Baird's tapir (*Tapirus bairdii*) with young.

usually defecate in the water at the river's
edge.

Tapirs are also good climbers, scrambling
up river banks and steep mountain sides
with great agility. They often follow the
same routes and may wear paths to standing water. They mark their territories and
daily routes with urine, as do rhinos. Tapirs
walk with their nose close to the ground,
probably to recognize their whereabouts
and to detect the scent of other tapirs and
predators. If threatened, they crash off into
the bush or defend themselves by biting.
Their main predators are big cats—the
jaguar in the New World and the leopard
and tiger in Asia. Bears sometimes prey on
Mountain tapirs and caymans will attack
young animals in the water.

Breeding occurs throughout the year, and
females appear to be sexually receptive
every two months or so. Mating is preceded
by a noisy courtship, with the participants
giving high-pitched squeals. Male and
female stand head to tail, sniffing their
partner's sexual parts and moving round in
a circle at increasing speed. They nip each
other's feet, ears and flanks (as do horses
and zebras) and prod their mate's belly with
their trunks.

Just before they give birth, pregnant
mothers seek a secure lair in which to bear
the single young (twins are rare). A female
in her prime can produce an infant every 18
months. Whatever the species, all newborn
tapirs have reddish-brown coats dappled
with white spots and stripes, excellent
camouflage in the mottled light and shade of
the jungle undergrowth. At two months old,
the pattern begins to fade, and by six months
the youngster has assumed the adult coat.
Youngsters stay with the mother till they
are well grown, but at 6–8 months they may
begin to travel independently; they are not
sexually mature for another two to three
years.

Although tapirs have survived for millions of years, their future is by no means
secure. They are hunted extensively for
food, sport and for their thick skins, which
provide good-quality leather, much prized
for whips and bridles. By far the greatest
threat is habitat destruction caused by log-
ging, clearing land for agriculture or man-
made developments such as the recent flooding of huge areas of forest for a new dam on
the Paraguay/Brazil border. The tapirs' best
hope for survival is in forest reserves like
Taman Negara in Malaya, but even here
there are pressures: part of Taman Negara is
threatened with flooding as part of a hydroelectric project.   KMack

RHINOCEROSES

Family: Rhinocerotidae [*]
Five species in 4 genera.
Order: Perissodactyla.
Distribution: Africa and tropical Asia.

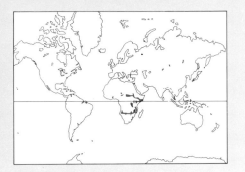

Size: head-body length from 98–124in
(250–315cm) in the Sumatran rhino to
146–158in (370–400cm) in the White rhino.
Weight from 1,765lb (800kg) in the Sumatran
rhino to 5,070lb (2,300kg) in the White rhino.

[*] CITES listed.

To many people, rhinoceroses with their massive size, bare skin and grotesque appearance are reminiscent of the reptilian dinosaurs which were the dominant large animals of the world between 250 million and 100 million years ago. Though rhinos are certainly not reptiles, it is true that they are relicts from the past. Rhinos of various forms were far more abundant and diverse during the Tertiary era (40–2 million years ago), but in Europe the Woolly rhinoceros survived until the last Ice Age (about 15,000 years ago).

While the extinct rhinos varied in their possession of horns and in the arrangement of these horns, they were generally large. Together with the elephants and hippopotamuses, they represent a life-form which was much more abundant and diverse in the past: that of the giant plant-feeding animals, or "megaherbivores". Of the surviving species, two are on the brink of extinction, while the other three are becoming increasingly threatened.

The name "rhinoceros" derives from the distinctive horns on the snout. Unlike those of cattle, sheep and antelopes, rhino horns have no bony core: they consist merely of an aggregation of keratin fibers perched on a roughened area on the skull. Both African species and the Sumatran rhinoceros have two horns in tandem, with the front one generally the largest, while the Indian and Javan rhinos have only a single horn on the end of the nose. Rhinos have short stout limbs to support their massive weight. The three toes on each foot give their tracks a characteristic "ace-of-clubs" appearance. The Indian rhino has an armor-plated look, produced by the prominent folds on its skin and its lumpy surface. The White rhino has a prominent hump on the back of its neck, containing the ligament supporting the weight of its massive head. In both the White rhino and the Indian rhino, adult males are notably larger than females, while in the other rhino species both sexes are of similar size. The Black rhino has a prehensile upper lip for grasping the branch ends of woody plants, while the White rhino has a lengthened skull and broad lips for grazing the short grasses that it favors. In color, the two species are not notably different, and the popular names most probably arose from the local soil color tinting the first specimens seen.

Rhinos have poor vision, and are unable to detect a motionless person at a distance of more than 100ft (30m). They eyes are placed on either side of the head, so that to see straight in front the animals peer first with one eye, then with the other. Their hearing is good, their tubular ears swiveling to pick up the quietest sounds. However, it is their sense of smell upon which they mostly rely for knowledge of their surroundings: the volume of the olfactory passages in the snout exceeds that of the brain! When undisturbed, rhinos can sometimes be noisy animals: a variety of snorts, puffing sounds, roars, squeals, shrieks and honks have been described for various species.

A further peculiarity of rhinos is that, as in elephants, the testes do not descend into a scrotum. The penis, when retracted, points backwards so that the urine is directed to the rear by both sexes. Females possess two teats located between the hindlegs.

The five surviving species of rhinos fall into three distinct subfamilies which are only distantly related to one another. The Sumatran rhino is the only surviving member of the Dicerorhinae, which also included the extinct Woolly rhino and other Eurasian species. The Sumatran rhino itself is little different from forms which existed 40 million years ago. The Asian one-horned rhinos (Rhinocerinae) have an evolutionary

▲ **Rhino in repose.** The massive horns of this placid Black rhino in Amboseli National Park, Kenya, pose no threat to the oxpecker perched on its upper jaw. Oxpeckers remove parasites from the rhinos' skin and also take insects stirred up by their dust baths.

◄ **Charging Black rhino.** Black rhinos are more aggressive than White rhinos.

► **An Indian rhino** in characteristic habitat: tall swampy grassland.

history extending back to Oligocene deposits in India; the Javan rhino is more primitive than the Indian rhino, having changed little over the past 10 million years. The African two-horned rhinos evolved independently in Africa. The White rhino is an offshoot of the same stock as the Black rhino, having diverged during the course of the Pliocene (about 3 million years ago).

All rhinoceroses are herbivores dependent on plant foliage, and they need a large daily intake of food to support their great bulk. Because of their large size and hindgut fermentation, they can tolerate relatively high contents of fiber in their diet, but they prefer more nutritious leafy material when available. Both African species of rhino have lost their front teeth entirely; although Asian species retain incisor teeth and the Sumatran rhino canines too, these are modified for fighting rather than for food gathering. The broad lips of White rhinos give them a large area of bite, enabling them to obtain an adequate rate of intake from the short grass areas that they favor for much of the year. Black rhinos use their prehensile upper lips to increase the amount of food they gather per bite from woody plants. Indian rhinos use a prehensile upper lip to gather tall grasses and shrubs, but can fold the tip away when feeding on short grasses; woody browse comprises about 20 percent of their diet during the winter period. Both the Javan and Sumatran rhinos are entirely browsers, often breaking down saplings to feed on leaves and shoot ends. They also include certain fruits in their diet, as also do African Black rhinos and, to a lesser extent, Indian rhinos.

All rhinos are basically dependent upon water, drinking almost daily at small pools or rivers when these are readily available. But under arid conditions, both African species can survive for periods of 4–5 days between water-hole visits. Rhinos are also dependent on water-holes for wallowing, Indian rhinos in particular spend long periods lying in water, while the African species more commonly roll over to acquire a mud coat. While the water may provide some cooling, the mud coating probably serves mainly to give protection against biting flies (despite the thick hide, blood vessels lie just under the thin outer layer).

For large, long-lived mammals like rhinos, life-history processes tend to be protracted. Female White rhinos and Indian rhinos undergo their first sexual cycles at about 5 years of age and bear their first calves at 6–8 years. In the smaller Black rhino, females breed about a year younger

than these ages. A single birth is the rule. Intervals between successive offspring can be as short as 22 months, but more usually vary between 2 and 4 years in natural populations of these three species. The babies are relatively small at birth, weighing only about 4 percent of the mother's weight—about 143lb (65kg) in the case of the White rhino and Indian rhino and 88lb (40kg) in the Black rhino. Females seek seclusion from other rhinos around the time of the birth. White rhino calves can follow the mother about three days after birth. Indian rhino mothers sometimes move away for up to 2,600ft (800m), leaving calves lying alone. Calves of both Indian and White rhinos tend to run in front of the mother, while those of Black rhinos usually run behind. White rhino mothers stand protectively over their offspring should danger threaten.

Males first become sexually potent at about 7–8 years of age in the wild; but they are prevented from breeding by social factors until they can claim their first territories or dominant status at an age of about 10 years.

Births may take place in any month of the year. In the African rhinos, conceptions tend to peak during the rains so that a birth peak occurs from the end of the rainy season through to the middle of the dry season.

Rhino are basically solitary, except for the association between a mother and her most recent offspring, which usually ends shortly before the birth of the next offspring. In White rhinos, and to a lesser extent in Indian rhinos, immature animals pair up or occasionally form larger groups. The White rhino is the most sociable of the five species, and females lacking calves sometimes join up, while such females also accept the company of one or more immature animals. In this way, persistent groups numbering up to seven individuals may be formed. Larger temporary aggregations may be found around resting areas or favored feeding areas. Adult males of all species remain solitary, apart from temporary associations with females in heat (see pp56–57).

In White rhinos and Indian rhinos, females move over home ranges covering 3.5–5.8sq mi (9–15sq km), with temporary extensions when food and water supplies run out. The home ranges of Black rhino females vary from about 1.2sq mi (3sq km) in forest patches to nearly 35sq mi (90sq km) in arid regions. Female home ranges in all species overlap extensively and there is no indication of territoriality among females. White rhino females commonly

▶ **Species of rhinoceros.** (1) Indian rhinoceros (*Rhinoceros unicornis*). (2) Sumatran rhinoceros (*Dicerorhinus sumatrensis*). (3) White rhinoceros (*Ceratotherium simum*). (4) Javan rhinoceros (*Rhinoceros sondaiacus*). (5) Black rhinoceros (*Diceros bicornis*).

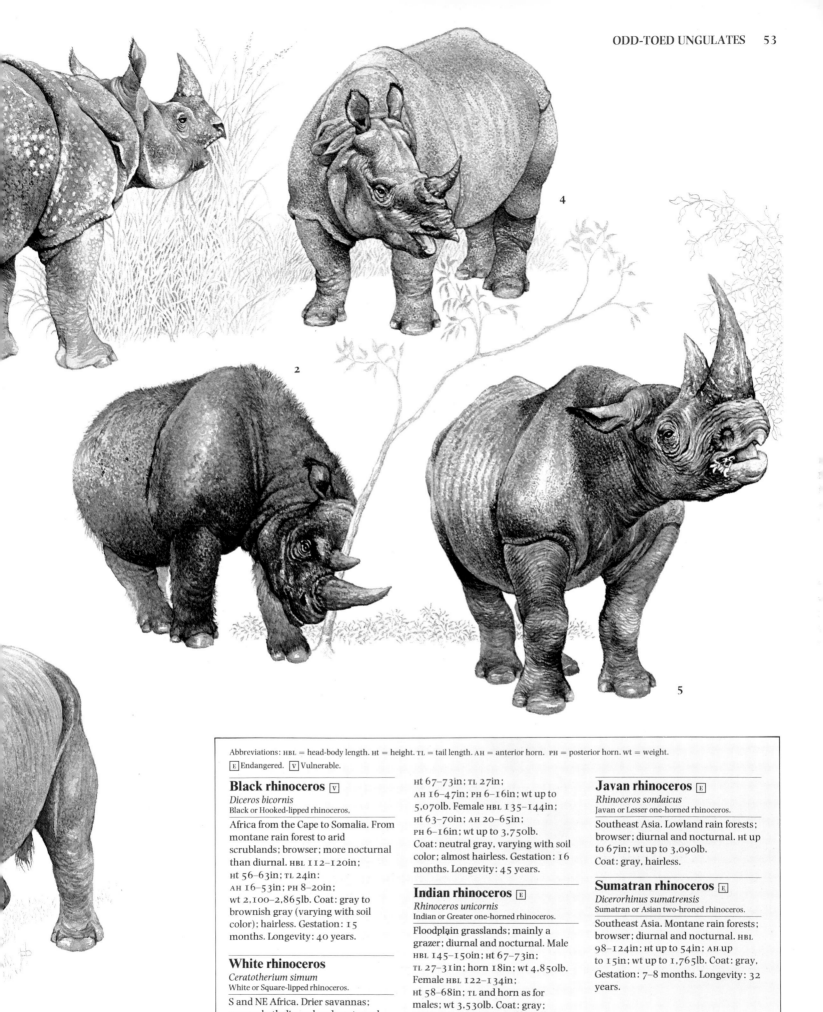

Abbreviations: HBL = head-body length. HT = height. TL = tail length. AH = anterior horn. PH = posterior horn. wt = weight.
E Endangered. V Vulnerable.

Black rhinoceros V

Diceros bicornis
Black or Hooked-lipped rhinoceros.

Africa from the Cape to Somalia. From montane rain forest to arid scrublands; browser; more nocturnal than diurnal. HBL 112–120in; HT 56–63in; TL 24in: AH 16–53in; PH 8–20in; wt 2,100–2,865lb. Coat: gray to brownish gray (varying with soil color); hairless. Gestation: 15 months. Longevity: 40 years.

White rhinoceros

Ceratotherium simum
White or Square-lipped rhinoceros.

S and NE Africa. Drier savannas; grazer; both diurnal and nocturnal. Male HBL 145–160in;

HT 67–73in; TL 27in; AH 16–47in; PH 6–16in; wt up to 5,070lb. Female HBL 135–144in; HT 63–70in; AH 20–65in; PH 6–16in; wt up to 3,750lb. Coat: neutral gray, varying with soil color; almost hairless. Gestation: 16 months. Longevity: 45 years.

Indian rhinoceros E

Rhinoceros unicornis
Indian or Greater one-horned rhinoceros.

Floodplain grasslands; mainly a grazer; diurnal and nocturnal. Male HBL 145–150in; HT 67–73in; TL 27–31in; horn 18in; wt 4,850lb. Female HBL 122–134in; HT 58–68in; TL and horn as for males; wt 3,530lb. Coat: gray; hairless. Gestation: 16 months. Longevity: 45 years.

Javan rhinoceros E

Rhinoceros sondaicus
Javan or Lesser one-horned rhinoceros.

Southeast Asia. Lowland rain forests; browser; diurnal and nocturnal. HT up to 67in; wt up to 3,090lb. Coat: gray, hairless.

Sumatran rhinoceros E

Dicerorhinus sumatrensis
Sumatran or Asian two-hroned rhinoceros.

Southeast Asia. Montane rain forests; browser; diurnal and nocturnal. HBL 98–124in; HT up to 54in; AH up to 15in; wt up to 1,765lb. Coat: gray, Gestation: 7–8 months. Longevity: 32 years.

engage in friendly nose-to-nose meetings, but Indian rhino females generally respond aggressively to any close approach. However, subadults of both species approach adult females, calves and other immature animals for nose-to-nose meetings and sometimes playful wrestling matches.

Males of all species sometimes fight viciously, inflicting gaping wounds. Both African species fight by jabbing one another with upward blows of their front horns. In contrast, the Asian species attack by jabbing open-mouthed with their lower incisor tusks, or, in the case of the Sumatran rhino, with the lower canines.

Black rhinos have a reputation for unprovoked aggression, but very often their charges are merely blind rushes designed to get rid of the intruder. However, if a human or a vehicle should fail to get out of their way, they can inflict much damage with their horns. Indian rhinos also frequently respond with aggressive rushes when disturbed, and may occasionally attack the elephants used as observation platforms in some of the sanctuaries where they occur. However, rhinos invariably come off second best in any fight with an elephant.

In contrast, the White rhino is mild and inoffensive by nature, and despite its large size is easily frightened off. Very often, a group of White rhinos will stand in a defensive formation with their rumps pressed together, facing outwards in different directions. While this formation may be successful against carnivores such as lions and hyenas, it is useless against a human armed with a gun.

Rhinos have been under threat from man for a long time. The three Asian species suffered a great reduction in numbers and considerable contraction of their ranges during the last century because of the local demand for their products. Following the advent of guns in Africa, the southern White rhino was reduced to the brink of extinction before the end of the 19th century. There is little local use in Africa, and products were generally exported. The Black rhino was exterminated in the Cape soon after the arrival of white settlers, but elsewhere in Africa remained widespread and fairly abundant until recently. However, with escalating trade between African countries and Asia, Black rhino numbers declined precipitously in East, Central and West Africa during the 1970s.

The reason behind the recent declines is the rapid increase in the value of rhino horn. While ground rhino horn is used as an aphrodisiac in parts of North India, its main

use in China and neighboring countries of the Far East is as a fever reducing agent. It is also used for headaches, heart and liver trouble, and for skin diseases. Many other rhino products, including the hooves, blood and urine, have reputed medicinal value in the East (but not in Africa). Chemically, rhino horn is composed of keratin, the same protein which forms the basis of hooves, fingernails and the outer horny covering of cattle and antelope horns, and there is no pharmacological basis for these uses; whatever success is achieved is probably psychological.

However, it is the use of rhino horns to

▲ **Rhino companions.** Rhinos are rarely without a few oxpeckers (here, on the nose of the right-hand rhino) and Cattle egrets in attendance.

▶ **Rhino horns** at Tsavo National Park, Kenya. The rhinos' most distinctive feature could also prove to be their downfall, as demand continues for their use as aphrodisiacs, as other medicinal agents and as dagger handles.

about 15,000 animals. The Sumatran rhino is now restricted to perhaps 150 individuals scattered throughout Sumatra, Malaya, Thailand and Burma. The Javan rhino, formerly widely distributed from India and China southwards through Indonesia, is now confined to a remnant of 50 in the Udjong Kulon Reserve in western Java. The Indian rhino is restricted to a few reserves in Assam, west Bengal, and Nepal, with a total population of about 1,500.

The situation of the White rhino is very different. The White rhino had a strange distribution, occurring in southern Africa south of the Zambezi River, and then again in northeastern Africa west of the Nile. The northern race has suffered severely from poaching in recent years, and has declined to perhaps a few hundred individuals in Zaire and the Sudan. The southern population was almost exterminated during the last century, but effective protection after 1920 resulted in a steady increase in the sole surviving population in the Umfolozi Game Reserve. By the mid-1960s their numbers had risen from perhaps 200 to nearly 2,000. As a result, the famous "Operation Rhino" was initiated by the Natal Parks Board to capture White rhinos alive for restocking other parts of their former range. This proved so successful that the species could be removed from the endangered list. By 1982, 1,200 White rhinos still remained in the Hluhluwe-Umfolozi Reserve in South Africa, and an equal number in other conservation areas in southern Africa. In fact, the main threat is posed by habitat deterioration due to the high densities attained by the species in the Umfolozi Game Reserve. To help provide outlets for the surplus animals which still need to be removed annually, some old males are sold to safari operators to be shot later by licensed hunters.

This contrasting situation is the source of an embarrassing conflict in conservation circles. While international cooperation is being sought to stop illegal hunting of rhinos through much of Africa, White rhinos, once the most endangered species, can be hunted legally in South Africa for legitimate reasons. While conservationists debate the most effective action, the situation is rapidly becoming desperate for rhinos in most regions of their occurrence. These hulking but simple-minded creatures are ill-adapted to cope with modern man armed with sophisticated weapons, and unless illegal hunting is controlled, rhinos may no longer be around by the turn of the century.

make handles for the "jambia" daggers traditionally worn by men in North Yemen as a sign of status that is mainly responsible for the recent rise in prices. Between 1969 and 1977 horns representing the deaths of nearly 8,000 rhinos were imported into North Yemen alone. The increase in the demand for rhino horn can be attributed to the fivefold increase in per capita income in Yemen as a result of oil wealth in the region.

The bulk of the pressure from the trade in rhino horns falls on the Black rhino in Africa and the Sumatran rhino in Asia. The Black rhino is still the most abundant and widespread species, but has been reduced to

NO-S

Horn to Horn
Territoriality, dominance and breeding in rhinos

Two male White rhinos approach each other to stare silently, horn to horn, then back away to wipe their horns on the ground. This ritual confrontation is repeated many times for perhaps up to an hour, before the males move apart to return to the hearts of their domains. For the point at which this ceremony takes place is the common boundary between their respective territories.

Territory holders also exhibit specialized techniques of defecation and urination, which may serve to scent mark the territories. The droppings are deposited at fixed dungheaps or middens, and are scattered by backwardly directed kicking movements. Especially large dungheaps, with prominent hollows developed by the kicking action, are located in border regions. The urine is ejected in a powerful aerosol spray, and urination is commonly preceded by wiping the horn on the ground then scraping over the site with the legs. Territory holders spray-urinate particularly frequently while patrolling boundary regions.

In White rhinos, access by males to receptive females is controlled by the strict territorial system. Prime breeding males occupy mutually exclusive areas covering 200–650 acres (80–260ha). These males form consort attachments to any females coming into heat that they encounter, and endeavor to confine such females within the territory for 1–2 weeks, until the latter are ready for mating. However, if the female should happen to cross into a neighboring territory, the male does not follow and the next-door male joins her.

Within territories, one or more subordinate males may be resident. Subordinate males do not spray their urine or scatter their dung, and they do not consort with females. When confronted by the territory holder, a subordinate male stands defensively uttering loud roars and shrieks. Females use similar roars to warn off males that approach too near. Generally, confrontations are brief, but if the subordinate male is an intruder from another territory a more prolonged and tense confrontation ensues, which may develop into a fight.

A defeated territory holder ceases spray-urination and dung scattering and takes on the status of subordinate male. Territory holders outside their own territories, on their way to and from water, also do not spray-urinate until they regain their own territories. If a territory holder is confronted by another male on a distant territory, he adopts the submissive stance and roars of a subordinate male. However, on a neighboring territory he maintains a dominant posture, but backs away steadily towards his own territory.

These behavior patterns signify a relationship whereby each territory holder is supremely dominant within the spatial confines of his own territory. This dominance gives him the opportunity to court and mate with any receptive female encountered there without interference from other males. Males nearing maturity, and deposed territory holders, choose to settle within a particular territory where the owner eventually becomes habituated to the presence of the additional male, providing that he displays subordinacy whenever challenged. While thus temporarily foregoing mating opportunities, subordinate males may gain strength to enable them at a later stage to challenge successfully for the status of territory holder in a nearby territory.

In a relatively high density population in the Hluhluwe Reserve, Black rhino breeding males occupy mutually exclusive home

▲ **Rhino confrontation.** A dominant White rhino confronts two subordinate rhinos on his territory. Subordinate males are tolerated by the dominant male provided they behave in a suitably submissive fashion.

◄ **Mating in rhinos** TOP can be a prolonged business, with several hours of foreplay, and copulations often lasting for one hour.

◄ **Rhino ritual.** In their confrontations, rhinos repeat the same gestures over and over before one concedes: (1) horns forced against each other; (2) wiping the horns on the ground. (3) A dominant male proclaims his mastery by spray-urinating, while the subordinate male retreats. Only dominant males spray-urinate.

areas which are shared by non-breeding males. These areas cover 1.5sq mi (4sq km), and meetings between neighboring males are rarely witnessed. In other Black rhino populations, the home ranges of males overlap and no clear evidence for territoriality has been found. Some males emit their urine in the form of a backwardly directed spray, but both males and females scatter their dung. When a female is in heat, several males sometimes displace one another in succession, before one succeeds in mating. Horn jousting matches between male and female sometimes occur during courtship.

In Indian rhinos, males can be classified as "strong" or "weak," but rather than being discrete categories there seems to be a continuum between them. Strong males urinate in a powerful backwards jet, associate frequently with females, and only they copulate. Such males move over home ranges covering up to 2.3sq mi (6sq km), but these overlap with those of other strong males, and are also shared by weak males. However, neighboring strong males rarely fight one another, while strange males entering from elsewhere are viciously attacked. Fights between male and female, and prolonged and noisy chases covering distances of several miles, are features of courtship.

These differences in social system can be related to differences in the density and distribution of food resources. As short grass grazers, White rhinos build up local densities in excess of 12.5 animals per sq mi (5 per sq km), while the location of favorable feeding areas at particular seasons is relatively predictable. Indian rhinos achieve local densities nearly as high, but because they are dependent upon flood plain habitats the location of favorable feeding areas changes in an unpredictable way. This does not favor spatial localization by males. In addition, the vegetation is dense, so that males may be screened from sensory contact with one another even when quite close by. For browsers the density of accessible food is much lower than is the case for grazers, and Black rhinos rarely exceed local densities of 2.5 per sq mi (1 per sq km). As a result, individuals occupy fairly large ranges and seldom come into contact, so that there is less pressure for males to avoid potentially risky contacts with other powerful males. Thus the control of mating rights seems to be more fluid, with stronger males claiming females from weaker ones when they come into contact. It is possible that in low-density populations of White rhinos the territorial system would be far less strongly expressed.

EVEN-TOED UNGULATES

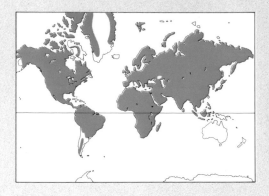

Order: Artiodactyla
One hundred and eighty-seven species in 76 genera and 10 families.
Distribution: worldwide, except Australia and Antarctica.

Habitat: very diverse.

Size: from head-body length 17–19in (44–48cm) in the Lesser mouse deer to 12–15ft (3.8–4.7m) in the giraffe. Weight from 4–5.7lb (1.7–2.6kg) in the Lesser mouse deer to 3,525–7,055lb (1,600–3,200kg) in the hippopotamus.

Pigs (family Suidae) – suids
Nine species in 5 genera.

Peccaries (family Tayassuidae) – tayassuids
Three species in 2 genera.

Hippopotamuses (family Hippopotamidae)
Two species in 2 genera.

Camels (family Camelidae) – camelids
Six species in 3 genera.

Chevrotains (family Tragulidae) – tragulids
Four species in 2 genera.

Musk deer (family Moschidae) – moschids
Three species in 1 genus.

Deer (family Cervidae) – cervids
Thirty-four species in 14 genera.

Giraffe (family Giraffidae)
Two species in 2 genera.

Bovids (family Bovidae)

Pronghorn (subfamily Antilocaprinae)
One species in 1 genus.

Wild cattle (subfamily Bovinae)
Twenty-three species in 8 genera.

Duikers (subfamily Cephalophinae)
Seventeen species in 2 genera.

Grazing antelope (subfamily Hippotraginae)
Twenty-four species in 11 genera.

THE even-toed ungulates, the Artiodactyla, are the most spectacular and diverse array of large, land-dwelling mammals alive today. Living in all habitats from rain forest to desert, from marshes to mountain crags, and found on all continents except Australasia and Antarctica, they dominated the mammal communities of the savannas where man's ancestors first arose. Indeed, they helped to mold the environment to which humans were adapted, and humans now control the environment in which dwindling communities of wild artiodactyls still survive. Early humans hunted them, and may have caused the extinction of some species. After the ice ages artiodactyls provided most of the large domesticated animals upon whose products and labor agricultural civilizations have depended.

Artiodactyls first appeared in the early Eocene (54 million years ago) in North America and Eurasia. They were at this time represented by small, short-legged animals, whose simple cusped teeth suggest a primate-like diet of soft herbage. By the late Eocene, the first members of the later, more advanced families were distinguishable, and by the Oligocene (35 million years ago) the three present-day suborders (Suina, Tylopoda, Ruminantia) were established. Piglike artiodactyls first appeared in Eurasia and Africa in the Oligocene, and ruminant artiodactyls made their first appearance in Africa in the early Miocene (25 million years ago).

The order Artiodactyla comprises two rather different types of animals, linked by similarities of anatomical structure. The suids (suborder Suina), including pigs and their relatives, are primarily omnivorous animals, retaining low-crowned cheek teeth with simple cusps, large tusk-like canines, short limbs, and four toes on the feet, although some lineages may be more "progressive" in the modification of these features. In contrast, ruminant artiodactyls, comprising the suborders Tylopoda (camels) and Ruminantia (deer, giraffes and bovids) are specialized herbivores that have evolved a multi-chambered stomach and adopted the habit of chewing the cud in order to digest fibrous herbage. They have cheek teeth with ridges rather than cusps, that may be high crowned, and show a progressive tendency to elongate the limbs and to reduce the number of functional toes to two from the five found in primitive mammals.

Living suids include the families Suidae (pigs), Tayassuidae (peccaries) and Hippopotamidae (hippos). Pigs are an exclusively Old World family. Peccaries are now confined to the New World, but were found in the Old World during the Tertiary. Members of both families are similar in having short legs, large heads and rooting snout, and are found in woodland, bushland and savannas in temperate and tropical habitats. In general, peccaries take a greater proportion of browse in their diet than the more omnivorous pigs, but some pigs, such as the African warthog, are specialist grazers. Hippos are today exclusively African tropical animals. The Pygmy hippo is a browser in the tropical forests, and the hippo is a semi-aquatic grazer. Hippos first appeared in the late Miocene (about 10 million years ago) in Africa, and radiated out into Eurasia in the

▶ **Artiodactyl horn types.** The extinct *Protoceratid* (1) had unbranched post-orbital horns and a curious forked nasal horn. *Dromomerycid* (2), another extinct species, had two unbranched supra-orbital horns and a single horn at the back of the skull. The Musk deer (3) is a primitive living artiodactyl with no horns and enlarged canines. The giraffe (4) has simple unbranched post-orbital horns in both male and female, covered by skin. The Roe deer (5) has branched post-orbital antlers (horn-like organs) which are shed yearly. Except in the reindeer only male deer have antlers. The pronghorn (6) has simple, unbranched, supra-orbital antlers with a keratinous cover that is shed annually. The female's antlers are smaller. Bovids, like the Common eland (7), have unbranched, post-orbital, keratin-covered horns that may be coiled or spiraled. Females in some species have horns.

Bone

Keratin

Deciduous bone

Deciduous keratin

Gazelles and Dwarf antelopes (subfamily Antilopinae)
Thirty species in 12 genera.

Goat antelopes (subfamily Caprinae)
Twenty-six species in 13 genera.

Skulls of Even-toed Ungulates

The skulls of pigs, like the babirusa, are characterized by a deep occiput (rear of the skull), an incomplete post-orbital bar, a short

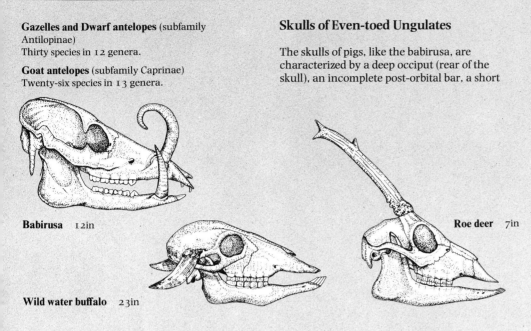

Babirusa 12in

Wild water buffalo 23in

Roe deer 7in

space between the front and back teeth, and the retention of large upper and lower canines. The teeth are usually low-crowned (except in some specialized grazers such as the warthog), simple cusped, and there is little tendency to molarize the premolars. The upper incisors are never entirely lost; the upper canines are characteristically curved upwards to form tusks in pigs, but point downwards in peccaries and hippos.

The skulls of ruminants, such as the Wild water buffalo and the Roe deer, are characterized by a shorter occiput, a complete post-orbital bar, a long gap between the front and back teeth and the tendency to reduce or completely lose the upper incisors. The molars may be low or high crowned, with the cusps coalesced into cutting ridges, and the premolars may be molarized. Ruminants resemble equids in the long face and posterior position of the orbits, but never develop the deep and massive lower jaw that characterizes horses.

Plio-Pleistocene (about 2 million years ago).

The only living tylopods are the camels and llamas (family Camelidae), which are low-level feeders that take a mixture of herbs and fresh grass. Camels are found today in arid steppes and deserts in Asia and North Africa, and llamas in high-altitude plains in South America, but until the Pleistocene (2 million years ago), camelids were an exclusively North American group. They first appeared in the late Eocene (40 million years ago), and only became extinct in North America 10,000 years ago.

The suborder Ruminantia can be further subdivided into the traguloids and the pecorans. Living traguloids are the chevrotains or mouse deer (family Tragulidae),

which are very small animals—2.2–4.4lb (1–2kg)—found today in the Old World tropical forests; they eat mainly soft browse which requires little fermentation. They lack horns, but possess saber-like upper canines. Traguloids first appeared in the late Eocene, and were widespread in North America as well as Eurasia in the early Tertiary. Chevrotains first appeared in the Miocene, and have always been restricted to the Old World.

The pecorans comprise those ruminants which possess bony horns, and the five living families are distinguished primarily by their horn type. The first pecorans were the hornless gelocids of the Eurasian Oligocene, and members of the horned families did not

begin to appear until the Miocene epoch.

Of the pecorans, giraffes (family Giraffidae) are high-level browsers, found today in African tropical forest (okapi) or savanna (giraffe), but they were also found in southern Eurasia during the Plio-Pleistocene. Deer (family Cervidae) are primarily small- to medium-sized browsers, with shorter legs and necks than giraffes, and are found today in temperate and tropical woodland in Eurasia, and North and South America. The Musk deer (family Moschidae), a small browser lacking horns and retaining large upper canines, is found today only in high-altitude forests in Asia, but in the Miocene there was a moderate radiation of moschids in North America, Europe and Africa.

The antelopes and cattle (family Bovidae) are today the most successful and diverse pecoran family, with a wide diversity of body sizes, forms and feeding adaptations. Most notably, they are the only family to have evolved medium- to large-sized open-habitat specialist grazers, such as the buffalo and the wildebeest. Bovids have been a primarily Old World group, reaching their peak of diversity in Africa, and they did not invade the New World until the Pleistocene. The pronghorn (subfamily Antilocaprinae) is also a surviving remnant of a once large radiation. Antilocaprids have always been exclusively North American, and the pronghorn is a medium-sized, long-legged, open-habitat, low-level browser on the western prairies. Here the pronghorn is regarded as a bovid, but some authorities consider it should be placed in its own family, suggesting a closer relationship with deer.

CJ/PJJ

EXTINCT FORMS

LIVING FORMS

WILD PIGS AND BOARS

Family: Suidae
Nine species in 5 genera.
Order: Artiodactyla.
Distribution: Europe, Asia, E Indies, Africa; introduced into N and S America, Australia, Tasmania, New Guinea and New Zealand.

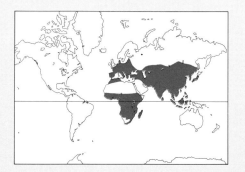

Size: head-body length from 23–26in (58–66cm) in the Pygmy hog to 51–83in (130–210cm) in the Giant forest hog. Weight from 13–20lb (6–9kg) in the Pygmy hog to 285–605lb (130–275kg) in the Giant forest hog.

▷ **Wild boar in a watery habitat.** Pigs are the most omnivorous of the hoofed mammals and like to root for food in moist soil.

▶ **The strange face** of the Giant forest hog, somewhat like that of a gorilla or a bat, but with tusks added!

▼ **Warthogs drinking.** Pigs usually forage in family groups like this.

WHAT wild pigs lack in grace and beauty they make up for in strength, adaptability and intelligence. They are admirably adapted to range the forests, thickets, woodlands and grasslands which they haunt in small bands, to duel for position and mates, to fend off predators and to enjoy a catholic diet. They are the most generalized of the living even-toed hoofed mammals (artiodactyls) and have a simple stomach, four toes on each foot and, in three of the genera, a full set of teeth.

The living wild pigs are medium-sized artiodactyls characterized by a large head, short neck and powerful, but agile, body with a coarse bristly coat. The eyes are small and the expressive ears fairly long. The prominent snout carries a distinctive set of tusks (lower canine teeth) and ends in a mobile, disk-like nose pierced by the nostrils. The structure of the snout, tusks and facial warts is intimately linked to diet, mode of feeding and fighting style. The tassled tail effectively swats flies and signals mood.

Key features of wild pigs are the well-developed, upturned canines, molars with rounded cusps, and a prenasal bone, which supports the nose. Pigs walk on the third and fourth digits of each foot, while the smaller second and fifth digits are usually clear of the ground. In warthogs, the first and second molars regress and disappear as the third molar elongates to fill the tooth row; in all pigs males are larger than females, with more pronounced tusks and warts. In the genera *Sus* and *Potamochoerus* the coat is striped at birth.

The senses of smell and hearing are well developed. Pigs are highly vocal, and family groups communicate incessantly by squeaks, chirrups and grunts. A loud grunt may herald alarm, while rhythmic grunts characterize the courtship chant which, in warthogs, sounds like the exhaust of a two-stroke engine.

Pigs usually forage in family parties. The Wild boar and bushpig feed on a wide range of plant species and parts (fungi, ferns, grasses, leaves, roots, bulbs and fruits) but also take insect larvae, small vertebrates (frogs and mice) and earthworms. They root for much of their food in litter and moist earth. The babirusa, the Giant forest hog and the warthog are more specialized herbivores. The babirusa feeds largely on fruits and grass; the Giant forest hog grazes predominantly in evergreen pastures and forest glades, seldom digging with its snout;

and warthogs feed almost entirely on grasses, plucking the growing tips with their unique incisors or with their lips. They take grass seeds at the end of the rains and during the dry season use the tough upper edge of the nose to scoop grass rhizomes out of the dry, sun-baked, savanna soils. Contrary to many popular accounts, warthogs seldom if ever use their tusks to dig out food.

Sexual maturity is attained at about 18 months, although males may only gain access to sows on reaching physical maturity at about 4 years old. Pigs may breed throughout the year in the moist tropics but in temperate and tropical areas of marked seasonality, mating occurs in the fall and sows farrow the following spring. Both the proportion of sows farrowing and the litter size (one or two in babirusa and up to 12 in the genus *Sus*) are influenced by prevailing environmental conditions. The young are born in a grass nest constructed by the mother or in a hole underground (warthog) and weigh between 18 and 32oz (500–900g). The piglets remain in the nest for about 10 days before following their mother. Each piglet has its own teat. Weaning occurs at about three months but young pigs remain with their mother in a closely knit family group until she is ready to farrow again. After farrowing in isolation, young sows may rejoin her. In this way, larger matriarchal herds or sounders are formed which may include several generations.

Abbreviations: HBL = head-body length. TL = tail length. wt = weight.

E Endangered.   V Vulnerable.   * CITES listed.

Pygmy hog E *
Sus salvanius

Himalayan foothills of Assam. Tall savanna grassland. Active daytime and twilight. Omnivorous.
HBL 23–26in; TL 1¼in; wt 13–22lb.
Coat: blackish-brown bristles on gray-brown skin; no facial warts.
Gestation: about 100 days.
Mammae: 3 pairs.
Longevity: 10–12 years.

Wild boar
Sus scrofa
Wild boar or Eurasian wild pig.

Europe, N Africa, Asia, Sumatra, Japan, Taiwan. Introduced into N America. Feral domestic pigs in Australia, New Zealand and N and S America. Broad-leaved woodlands and steppe. Active daytime and twilight.
HBL 35–71in; TL 12–16in; wt 110–440lb.
Coat: brownish-gray bristles; short dense winter coat; no facial warts.
Gestation: 115 days.
Mammae: 6 pairs.
Longevity: 15–20 years.

Javan warty pig
Sus verrucosus

Java, Madura and Bawean. Forest, lowland grasslands and swamps, probably forest. HBL 35–63in; TL ?; wt up to 410lb.
Coat: red or yellow hair with black tips; marked facial warts.
Gestation and longevity: probably as for Wild boar.

Bearded pig
Sus barbatus

Malaya, Sumatra and Borneo. Tropical forest, secondary forest and mangroves. Active daytime.
Coat: dark brown-gray with distinctive white beard on cheeks; marked facial warts.
Size, gestation and longevity: probably as for Wild boar.

Celebes wild pig
Sus celebensis

Sulawesi. Lowland and upland tropical forest. Smaller than Bearded pig. Well-developed facial warts.

Babirusa V
Babyrousa babyrussa

Sulawesi, Togian, Sulu and Buru Islands. Tropical forest. Active daytime. HBL 33–41in; TL 11–12in; wt up to 200lb.
Coat: sparse, short, white or gray bristles; no facial warts.
Gestation: 125–150 days.
Mammae: 1 pair.
Longevity: up to 24 years.

Bushpig
Potamochoerus porcus
Bushpig or Red river hog.

Subsaharan Africa and Madagascar. Forest, moist savanna woodlands and grasslands. Active day and night.
HBL 39–59in; TL 12–16in; wt 110–265lb.
Coat: brick red to gray bristles with white, often patterned, face and mane; facial warts in male.
Gestation: 127 days.
Mammae: 3 pairs.
Longevity: 10–15 years.

Giant forest hog
Hylochoerus meinertzhageni

Central African Congo basin, parts of West and East Africa. Tropical forest and intermediate zone between forest and grassland. Active daytime.
HBL 51–83in; wt 12–18in; wt 285–600lb;
Coat: brown and black bristles; facial warts.
Gestation: 149–154 days.
Mammae: 3 pairs.
Longevity: not known.

Warthog
Phacochoerus aethiopicus

Subsaharan Africa. Savanna woodland and grasslands. Active daytime. HBL 43–53in; TL about 16in; wt 110–245lb.
Coat: black or white bristles on gray skin; pronounced facial warts.
Gestation: 170–175 days.
Mammae: 2 pairs.
Longevity: 12–15 years.

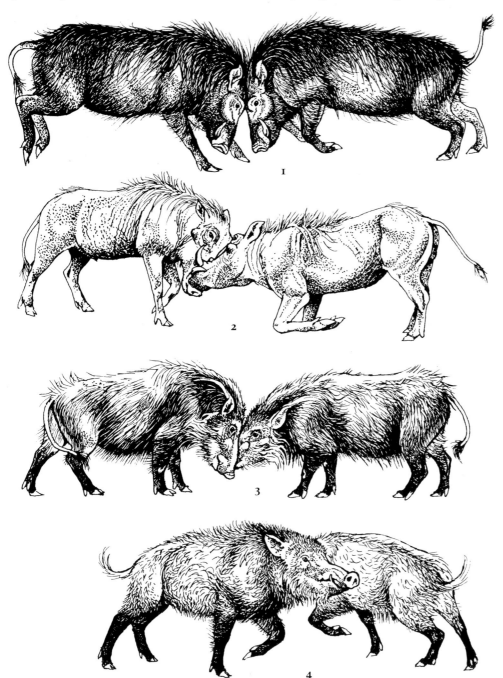

◀ **The warthog's tusks.** Although the upper tusks are more impressive, it is the smaller but sharper lower tusks that are the warthog's principal weapons.

▼ **Fighting styles of pigs.** The distinctive patterns of weapons and armor displayed by each species reflects the mode of combat. (1) In the Giant forest hog, contact is frontal and the top of the head is toughened. (2) In the warthog, contact is also frontal and the facial warts protect against the incurving tusks. (3) In the bushpig, snouts are crossed, sword-like, and are similarly protected by warts. (4) Wild boars slash at each other's shoulders, which are protected by thickened skin and matted hair.

In courtship, a chanting boar nudges the sow's flanks, sniffs her genital region, indulges in lateral displays and repeatedly attempts to rest his chin on her rump—a stimulus which causes a fully receptive sow to stand. The bushpig, warthog, and pigs of the genus *Sus*, and possibly other pigs, produce lip-gland pheromones which may be dispersed while chanting; *Sus* also produces a salivary foam. Mating may last 10 minutes and the spiral penis fits into a grooved cervix in which a plug forms after copulation.

The main social units are solitary boars, bachelor groups and matriarchal sounders comprising one or more adult sows with their young of various ages. Fragmentation of larger mother-daughter groups leads to kinship units or clans comprising a number of related sounders, with overlapping home ranges, which share feeding grounds, water holes, wallows, resting sites and sleeping dens. Wild pigs appear to be non-territorial, with home ranges of about 61 acres (25ha) in Pygmy hogs, 0.4–1.5sq mi (1–4sq km) in warthogs and 3.8–7.7sq mi (10–20sq km) in Wild boars. Home ranges are marked by lip glands, pre-orbital glands (warthog and Giant forest hog) or foot glands (bushpig). The mating system appears to be a roving dominance hierarchy among the males within a clan area. In northeast Borneo, the Bearded pig is noted for the large migrations it undertakes.

Habitat destruction threatens the Pygmy hog, and the Javan warty pig popupalation is critically low; they require both habitat protection and vigorous captive breeding programs. If the Javan warty pig is to be safe in the wild it requires translocation to protected islands which exclude the Wild boar as a competitor. The position of the African wild pigs and the widely distributed Wild boar is presently satisfactory.

Man has reared pigs for meat in Europe and Asia since neolithic times, and well-developed pig cultures still exist in parts of Indonesia. The Wild boar has been domesticated separately in Europe, India, China and Malaya. A domesticated form of the Celebes wild pig occurs in the Philippines, Sulawesi and neighboring islands.

The African wild pigs are symptomless carriers of African swine fever (ASF), a virus disease transmitted by the tampan, a soft-bodied tick, and lethal to domestic swine. ASF may acount for the absence of feral swine south of the Sahara. This has led to eradication campaigns against the bushpig, warthog and Giant forest hog, where the disease has threatened pig farming. African wild pigs also support blood-sucking tsetse flies, the carriers of trypanosomes responsible for sleeping sickness in man and ngana in domestic livestock. The Savanna tsetse fly is particularly partial to warthog, and elimination of warthogs and bushpigs has played an important part in controlling ngana, which covers an area of Africa equivalent to the USA.

Whereas the warthog is readily exterminated, the bushpig has taken advantage of agricultural development in many parts of Africa and is notorious for its nocturnal raids on cultivated crops. In America, Australia and New Zealand, feral pigs cause damage to lambs, cultivated crops and local plant communities.   DHMC

PECCARIES

Family: Tayassuidae
Three species in 2 genera.
Order: Artiodactyla.
Distribution: SW USA to N Argentina.

Collared peccary ⟨∗⟩
Tayassu tajacu
Collared peccary, báquiro, chacharo, javelina or javalí.
Distribution: SW USA to N Argentina.
Habitat: low and high dry tropical forest, wet tropical forest, chaparral and oak-grasslands; active mainly during day.
Size: head-body length 31–38in (80–98cm); shoulder height 12–16in (30–40cm); tail length 1.4–1.8in (3.5–4.5cm); weight 38–55lb (17–25kg). Coat: grizzled gray with dark grayish back; limbs blackish; whitish band diagonally from middle back to chest; juveniles russet colored and collar diffuse. Gestation: 142 days. Longevity: 8–10 years (21 years in captivity).

White-lipped peccary
Tayassu pecari
White-lipped peccary, pecari, careto, cachete blanco, cochino salvaje.
Distribution: SE Veracruz state, Mexico to N Argentina.
Habitat: high dry tropical forest, wet and rain tropical forest; active day and night.
Size: head-body length 39–48in (100–120cm); shoulder height 20–24in (50–60cm); tail length 1.6–2in (4–5cm); weight 55–88lb (25–40kg). Coat: grayish black to dark brown, with whitish bristles on lips, chin, throat and rump; juveniles uniform gray. Gestation: 158 days. Longevity: unknown.

Chacoan peccary ⟨v⟩
Catagonus wagneri
Chacoan or chaco peccary, pagua or tagua.
Distribution: Gran Chaco (N Argentina, SE Bolivia, W Paraguay).
Habitat: thorny forest with isolated islands of palm and grasses, active during day.
Size: head-body length 38–45in (96–117cm); shoulder height 21–25in (52–69cm); tail length 1.2–4in (3–10cm); weight 68–95lb (31–43kg). Coat: grizzled gray-brown, black and white, with black dorsal stripe and white collar; limbs blackish; juveniles tan and black with diffuse tan collar. Gestation: not known. Longevity: not known.

⟨v⟩ Vulnerable. ⟨∗⟩ CITES listed.

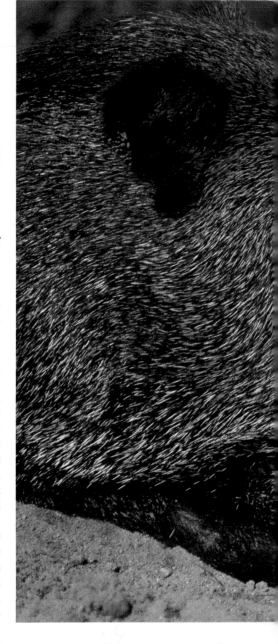

SOUTH American peasants claim " that the jaguar only kills peccaries that remain apart from the group." This saying has been verified, since it is now known that individual peccaries will confront a predator, at considerable risk to themselves, while other members of their group escape. Such altruistic anti-predator behavior is just one of the unusual features of the life-style of peccaries, gregarious inhabitants of the tropical and subtropical forests of the Americas. One of the species—the Chacoan peccary—was discovered by scientists as recently as 1975.

Peccaries are medium-sized pig-like animals with long, slender legs. The three species differ in color as well as in size: the Collared peccary has a white collar, the White-lipped peccary white lips, and the Chacoan peccary both the white collar and lips. Compared to the two species of *Tayassu*, the Chacoan peccary is slightly larger and has a bigger skull, its dental crowns are more developed, its eyes are located further to the rear of the head and its snout is longer—differences which justify placement in the separate genus *Catagonus*. Males and females are the same size in all species. Coat color is different between juveniles and adults in each species.

Both peccary genera evolved from the North American genus *Platygonus* from the upper Pliocene (about 5–2 million years ago). *Platygonus* colonized South America with the formation of the South American land bridge. *Catagonus* is more closely related to *Platygonus* than is *Tayassu*.

Peccaries are omnivorous, but feed preferentially on roots, seeds and fruits. Occasionally, they also eat insects and other invertebrates. Cacti figure largely in the diets of Chacoan and Collared peccaries. They have three compartments in their stomachs in which it is thought microbial flora digest cellulose, as occurs in ruminants. Their tusks are used to cut roots, which they consume avidly. Their jaw muscles are well developed and are strong enough for peccaries to crush tough seeds. The jaws of the White-lipped peccary are especially powerful. Jaw movement is up-and-down rather than gyratory, as in other hoofed mammals, Peccaries are known to lick and eat soil, possibly as a source of mineral nutrients.

Female Collared peccaries are sexually mature at 33–34 weeks, males at 46–47 weeks. Copulation lasts just a few seconds and is not preceded by any courtship displays. Females may mate with several males, with adult males establishing a hierarchy to prevent subordinates mating.

Unlike pigs, which are mostly solitary,

peccaries live in herds of 14–50 individuals in the Collared peccary, of 100 or more in the White-lipped peccary and of 2–10 in the Chacoan peccary. Each herd is subdivided into small family groups, which in the Collared peccary comprise 13–14, including males, females, females with young, and juveniles; females may outnumber males by 3 : 1. During the dry season, when food is not abundant, these groups often separate to

▲ **The canines of peccaries,** visible here in this Collared peccary, are short, sharp, and conventional compared to the exotic tusks of the related pigs. They have 38 teeth but the first premolars are reduced. Their coat is a distinctive feature, comprising tough, thick hairs 0.8–8.6in (2–22cm) long, the longest on the mane and rump. A conspicuous gland, up to 3 × 0.4in (7.5 × 1cm) in size, is present on the rump.

◄ **A young Collared peccary,** at one month old. Young remain dependent on their mothers for about 24 weeks. Lactation lasts 6–8 weeks, although the young eat solid food within 3–4 weeks. Both parents and other group members help to care for the young. In the face of threatened danger, parents and non-parents alike will shelter the young between their rear legs, and the young are allowed unrestricted access to high-quality or limited food.

search for food. Even when a herd is resting, groups tend to disperse to different parts of the site. Such groups are enduring and are particularly cohesive when juveniles are present.

Peccaries occupy stable territories, which are strongly defended from intruders. Territories of the Collared peccary vary in size from 74–690 acres (30–280ha) depending on vegetation type and quantity and distribution of food resources. Secretions from the rump gland are used by adults and subadults to mark tree trunks and rocks within the territory. The core area, which is preferentially used, is also marked by dung piles, one of which is used throughout the year and others only seasonally. These dung piles are important parts of the forest habitat since they contain large numbers of undigested seeds which readily germinate to rejuvenate the forest and disperse the trees.

Cohesion within groups is reinforced by mutual marking by a gland below the eyes—individuals stand side-by-side, turning their faces to each other and rubbing

each other's glands. Such marking is believed to aid recognition between all members of the group. In the Collared peccary, boisterous play and mutual grooming and scratching with snouts also often occur at the start of a day, before feeding.

Groups of Collared and Chacoan peccaries apparently have no leader, but when moving from one feeding site to another they follow whichever individual happens to be at the front of the line, usually one of the females in the group, followed by other adult females, young and juveniles.

Peccaries are vocal animals, and six types of vocalization can be distinguished in the Collared peccary. A cough-like grouping call by an adult male recalls dispersed individuals to the group. A repeated dry short "woof" is the alarm call and a "laughing" call is used in aggressive encounters between individuals. A clear nasal sound is produced when peccaries eat. Infants that have strayed indicate their distress by emitting a shrill clucking call which prompts the mother to seek them out. A repeated rasping sound, produced by chattering the teeth, is emitted by animals that are angry or annoyed. During squabbles, the neck and back bristles are often raised.

The main predators of peccaries are mountain lions and jaguars. Two forms of anti-predator behavior have been observed. In the Collared peccary, if the predator gets very close before detection, a herd or group disperses in all directions, to confuse the assailant, while emitting the alarm call. If young are present and the habitat is dense, one individual, usually a subadult of either sex, may approach the predator, in an act of apparent altruism, while others escape as the predator's attention is diverted. The outcome for the vigilant animal may be fatal. When resting, some males generally remain vigilant around the periphery of groups, and these are replaced by rested males during the period.

Clearance of forest for crops and pastures is reducing peccary habitat, and management plans are needed to conserve peccary populations. Indians and peasants hunt peccaries in the tropics for their meat; their gregarious nature and wide distribution make them readily accessible. The Chacoan peccary is the most susceptible to human disturbance as its distribution is the most restricted. Collared peccaries are often considered pests because the eat and destroy plantations of yucca, corn, watermelons and legumes, to the point where local campaigns have been mounted to exterminate them. HGC

HIPPOPOTAMUSES

Family Hippopotamidae
Two species in 2 genera.
Order: Artiodactyla.
Distribution: Africa.

Pigmy hippopotamus [v] [*]
Choeropsis liberiensis
Distribution: Liberia and Ivory Coast; a few also in Sierra Leone and Guinea.

Habitat: lowland forests and swamps.

Size: head-body length 4.9–5.7ft (1.5–1.75m); height 30–39in (75–100cm); weight 397–605lb (180–275kg).

Skin: slaty greenish black, shading to creamy gray on the lower part of the body.

Gestation: 190–210 days.

Longevity: about 35 years (42 recorded in captivity).

Hippopotamus [*]
Hippopotamus amphibius
Distribution: W, C, E and S Africa.

Habitat: short grasslands (night); wallows, rivers, lakes (day).

Size: head-body length 10.8–11.3ft (3.3–3.45m); height 4.6ft (1.4m); weight male 3,525–7,055lb (1,600–3,200kg), female 3,086lb (1,400kg).

Skin: upper part of body gray-brown to blue-black; lower part pinkish; albinos are bright pink.

Gestation: about 240 days.

Longevity: about 45 years (49 recorded in captivity).

[v] Vulnerable. [*] CITES listed.

▶ **Pigmy hippos,** showing the sleek skin which, like that of the hippopotamus, is covered in pores which exude a sticky fluid.

THE hippos are an example of the unusual phenomenon of a species pair, ie two closely related species that have adapted to different habitats (other examples of species pairs are Forest and Savanna elephants and buffalos). The smaller hippo occupies forest, the larger grassland. The hippo is also unusual for the way its daytime and nighttime activities, and feeding and breeding, take place in very different habitats: daytime activities in water, nighttime ones on land. This division of life zones is associated with a unique skin structure which causes a high rate of water loss when the animal is in the air and therefore makes it necessary for the hippo to live in water during the day.

The body of the hippo is barrel-shaped with short, stumpy legs. Its large head is adapted for an aquatic life, with eyes, ears and nostrils placed on the top of the head. It can submerge for up to five minutes. Its lips are up to 20in (50cm) broad and the jaws can open to 150 degrees wide. Its most characteristic vocalization is a series of staccato grunts.

The Pigmy hippopotamus resembles a young hippopotamus in size and anatomy, with proportionately longer legs and neck, smaller head and less prominent eyes which are placed to the side rather than on top of the head. It is less aquatic than the hippo-

potamus; the feet of the Pigmy hippo are less webbed and the toes more free.

The skin of the hippo is smooth, with a thick dermis and an epidermis with unusually thin, smooth and compact surface layer—a unique adaptation to aquatic life. Because of the very thin cornified layer, the skin seems to act as a wick, allowing the transfer of water, and the rate of water loss through the epidermis in dry air is several times greater than in other mammals, losing about 12mg from every 0.7sq in every 10 minutes—about 3–5 times the rate in man under similar conditions. The Pigmy hippo has a similar rate. This necessitates resort to a very humid or aquatic habitat during the day, otherwise the animal would rapidly dehydrate. Another characteristic of the skin is the absence of sebaceous glands or true temperature-regulating sweat glands. Instead there are glands below the skin which secrete a pink fluid which dries to form a lacquer on the surface. This gave rise in the past to the idea that the hippo "sweats blood." The "sweat" is very viscous, and highly alkaline. The red coloration is probably due to skin pigments similar to those which cause tanning in man. The pigment is not penetrated by ultraviolet radiation and

protects the skin from sunburn. It may also have a protective function against infection, because even large wounds are always clean and free from pus, despite the filthy water of wallows and lakes.

The fossil record shows that there were once several species of hippos. Only two have survived, of which the Pigmy hippo is not only the smaller but more primitive. It probably bears the closer resemblance to the ancestral hippo, which, like the Pigmy, lived in forests. A speculative view of the evolution of the hippos envisages their origin in a small forest-living form similar to the Pigmy hippo. In the course of time, forests were replaced by grasslands and the ancestral hippo eventually moved out of the forest to occupy gallery forest or thickets along the rivers and lakes by day, grazing in open grasslands by night. This gave access to the abundant food supply of the tropical savanna grassland and was accompanied by development of much larger size, but the animal's unusual physiology required terrestrial feeding at night and resort to water during the day. There was a behavioral adaptation to the grassland environment rather than a physiological one and the consequent penalty is a limited geographical

▲ **Looking every inch an aquatic mammal,** a bull hippo stalks a river bed. The hippo is forced to spend the daytime in water because the skin loses water at a very high rate in air.

▼ **Hippos "blood."** The red exudate from the skin glands of hippos led to an early belief that they "sweated blood." The secretion is a protection against the sun and probably also against infections.

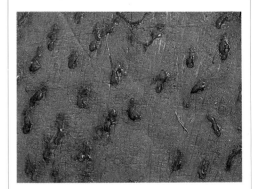

▷ **Nothing like mud!** OVERLEAF By using muddly wallows up to 6.2mi (10km) from water, hippos can extend their grazing range further inland.

range, restricted to the vicinity of water.

The hippos feed almost exclusively on terrestrial vegetation. Very little is known of the food of the Pigmy hippo, but it probably feeds on roots, grasses, shoots and fruits found on the forest floor and in swamps. The hippopotamus feeds almost exclusively on a number of species of short grasses, which are cropped by a plucking motion of the broad lips. This means that shallow-rooted grasses are selectively removed and at high population densities overgrazing and soil erosion ensues (see box overleaf).

Feeding is compressed into five or six hours of the night; the rest of the time is spent in the water. Food is digested in ruminant-like manner in compartmented stomachs, but the process seems inefficient. However, daily food requirement only averages 88lb (40kg) a night, equal to a dry weight of about $1-1\frac{1}{2}$ percent of body weight, compared with $2\frac{1}{2}$ percent for other hoofed mammals. This is probably because the way of life minimizes energy expenditure: the active feeding period is short and for the remainder of the 24 hours the hippo is resting or more or less inactive, in a supportive, stable, warm medium. Even energy requirements for muscle tone are reduced: the hippo most often opens its mouth to yawn!

Mating is correlated with dry seasons when the population is most concentrated. Both species prefer to mate in water, the female partly or wholly submerged, lifting her head from time to time to breathe. The proportion of females pregnant in one season can vary between only 6 percent in dry years and 37 percent in wetter periods. Sexual maturity is attained at an average of 7 years in the male (range 4–11 years) and 9 years in the female (range 7–15 years). Social maturity in the male is reached at about 20 years. In captivity, sexual maturity is attained in the Pigmy hippo at about 4–5 years.

The birth season occurs during the rains, resulting in a single peak of births in southern Africa and a double peak in East Africa. To give birth, a female leaves her group and a single young is born on land or in shallow water, weighing about 93lb (42kg). Occasionally, it is born underwater and has to paddle to the surface to take its first breath. Twins occur in less than 1 percent of pregnancies. The mother is very protective, and when she and her offspring rejoin the herd 10–14 days after birth the young may lie across the mother's back in deeper water. Suckling occurs on land, and in or under water, and the duration of lactation is about

8 months. Natural mortality in the first year is about 45 percent, reducing to 15 percent in the second year, then 4 percent per annum to about 30 years, subsequently increasing.

The average group size is 10–15 in hippos, the maximum 150 or so, according to locality and population density, but only 1–3 in the Pigmy hippo. In the latter, the triads are usually male, female and calf; their social behavior is almost unknown. The hippo, however, is either solitary (usually male individuals) or found in nursery groups and in bachelor groups. The solitary males are often territorial; other males maintain territories containing the nursery groups of females and young. In a stable lake environment some bulls maintain territories for at least eight years; in a less stable river environment most territories change ownership after a few months, but some are maintained for at least four years. Hippo densities in lake and river are respectively 7 and 33 per 330ft (100m) of shore. The territories are 550–1,000ft (250–500m) and 110–220ft (50–100m) in length respectively. Territorial bulls have exclusive mating rights in their territories but tolerate other bulls if they behave submissively. Non-territorial bulls do not breed. The frequency of fighting is reduced by an array of ritualized threat and intimidation displays, but when fights do occur they may last up to one and a half hours and are potentially lethal. The razor-sharp lower canines are up to 20in (50cm) long; the canine teeth have a combined weight of up to 2.4lb (1.1kg) in females and 4.6lb (2.1kg) in males. Defense is by the well-developed dermal shield thickest along the sides. Adult males can receive severe wounds and are often seen heavily scarred, but the skin possesses a remarkable ability to heal. Displays include mouth opening, dung scattering, forward rushes and dives, rearing up and splashing down, sudden emergence, throwing water (using the mouth as a bucket), blowing water through the nostrils, and vocalizations, especially a series of loud staccato grunts. Submissive behavior is a lowering of the head and body. Male territorial boundary meetings are ritualized: they stop, stare, present their rear ends and defecate while shaking the short, vertically flattened tail to spread the dung, before walking back to the center of their territories.

In the evenings, the aquatic groups break up and the animals go ashore, either singly or as females with their calves. They walk along regular trails marked by dung piles that serve as scent markers for orientation at

Conservation and Management

The highest density hippo populations in the world are found in East and Central Africa in the lakes and rivers of the western Rift Valley, particularly around Lakes Edward and George. Maximum grazing densities recorded in the Queen Elizabeth National Park were formerly over 81 per sq mi (31 per sq km), equivalent to 177,000lb per sq mi (31,000kg per sq km) of hippo alone, not counting other grazing and browsing hoofed mammals. They create an energy and nutrient sink from land to water by removing grass and defecating in the lakes. This promotes fish production by fertilizing the water.

Overgrazing within 1.8mi (3km) of the water's edge led to erosion gulleys and reduction of scrub ground cover. The Uganda National Parks Trustees initiated a hippo management culling program in 1957, the first such management for any species in Africa. Subsequently, from 1962–1966, an experimental management scheme was operated, with the aim of maintaining a range of different hippo grazing densities in a number of management areas from zero to 60 per sq mi (23 per sq km) while climate, habitat changes and populations of other herbivores were monitored to establish the influence of different hippo densities. About 1,000 hippos were shot annually over 5 years, and the meat was sold to local people. In one of the experimental areas there were 90 hippos in 7sq mi (4.4 sq km) in 1957, and it was bare ground except for scattered thickets. By 1963, after the near elimination of hippos, the grass had returned and erosion was halted, while average numbers of other grazers increased from 40 to 179 and the total hoofed mammal biomass increased by 7.7 percent; this trend continued until 1967, when management culling ceased and reversion towards the previous situation began. Subsequently, the

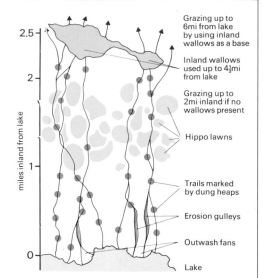

Grazing up to 6mi from lake by using inland wallows as a base

Inland wallows used up to 4½mi from lake

Grazing up to 2mi inland if no wallows present

Hippo lawns

Trails marked by dung heaps

Erosion gulleys

Outwash fans

Lake

breakdown of law and order in Uganda interrupted research and led to serious poaching.

As the hippo population was reduced, the mean age of sexual maturity declined from 12 years to under 10 years, and the proportion of calves increased from 6 percent to 14 percent. The influence of climatic change is difficult to establish in the short term, but for the rainfall levels of the 1960s the predicted grazing density for hippo in this area of Uganda was about 8 per sq km (21 per sq mi) if soil erosion was to be controlled, and if vegetation cover and variety, and herbivore numbers and diversity were to be maintained.

This work indicated the optimal rate of culling in one locality. However, the optimum level of hippo grazing density varies according to soil type, vegetation composition and climate; it needs to be established by management research for any particular area. Management culling remains a controversial issue, but almost all current knowledge about hippo populations is the result of such schemes.

▲ **Lunging hippos** contest a river territory. Hippo fights can be lethal, and dominant males can maintain territories for up to eight years.

◀ **Young hippos** ride piggy-back on their mothers in deeper water. Birth sometimes takes place in the water and some suckling is carried out there too.

night and have no territorial function. The paths have been measured to 1.7mi (2.8km) long and branch at the inland end, leading to the hippo lawns, which are discrete short-grass grazing areas. The animals share a common grazing range, where they feed; they then return to the water along the tracks during the night or early morning. Observations of identifiable individuals indicate that the composition of aquatic groups can remain fairly constant over a period of months, although they are very unstable compared with, for example, elephant family groups. The Pigmy hippo has similarly marked trails on the thinly vegetated forest floor or tunnel-like paths through denser vegetation.

Both species occupy wallows, which may be very large in the case of the hippopotamus; smaller wallows are occupied predominantly by males. This allows the grazing range to be extended inland up to 6.2mi (10km) or more. High hippo densities, in some areas 80 per sq mile have led to overgrazing and management culling (see box).

Both species play host to several animals. An association with a cyprinid fish, *Labio velifer*, has been described: it is sometimes temporarily attached and may graze on algae or other deposits on the hippo skin. Terrapins and young crocodiles bask on the backs of hippos, Hammer-headed storks and cattle egrets use them as a perch for fishing, and other birds, including oxpeckers, are associated. A parasitic fluke, *Oculotrema hippopotami*, is uniquely found attached to 90 percent of hippos' eyes; on average 8, but up to 41 have been found on one hippo.

RML

CAMELS AND LLAMAS

Family: Camelidae
Six species in 3 genera (including 3 domesticated species).
Order: Artiodactyla.
Distribution: SW Asia, N Africa, Mongolia, Andes.

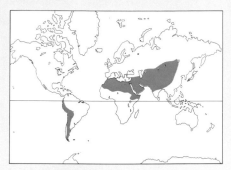

Size: ranges from shoulder height 34–38in (86–96cm); weight 99–121lb (45–55kg) in the vicuna to height at hump 75–91in (190–230cm), weight 1,000–1,450lb (450–650kg) in the dromedary and Bactrian camel.

▼ **Well-wrapped up** in its winter coat, the Bactrian camel has an ungainly, shaggy appearance. In spring, when the winter coat is shed, they look even more tattered.

MEMBERS of the camel family are among the principal large mammalian herbivores of arid habitats and they make a crucial contribution to man's existence and survival in desert environments. The domesticated one-humped dromedary of southwestern Asia and North Africa, and the two-humped Bactrian camel, which is still found wild in the Mongolian steppes, are well-known, but there are also four species in the New World: the "lamoids" or "cameloids." Of these, the vicuna and guanaco are wild and the llama and alpaca are domesticated.

The camel differs from other hoofed mammals in that the body load rests not on the hoofs but on the sole-pads, and only the front ends of the hoofs touch the ground. The South American cameloids are adapted to mountainous terrain, and the pads of their toes, which are not as wide as those of camels, are movable to assist them on rocky trails and gravel slopes. The split upper lip, the long curved neck and the lack of tensor skin between thigh and body, so that the legs look very long, are characteristic of camels. They move at an ambling pace. Camels are unique among mammals in having elliptical blood corpuscles. They all have an isolated upper incisor, which in the males of the lamoids is hooked and sharp-edged, like the tusk-like canines found in both jaws. Camels have horny callosities on the chest and leg joints.

The camelids first appeared in the late Eocene (among the earliest of the even-toed hoofed mammals). The camel family originated and evolved during 40–45 million years in North America, with key dispersals to South America and Asia occurring only 2–3 million years ago.

The relationships between the South American lamoids are confused. Fertile offspring are produced by all possible pure and hybrid matches between the four and they all have the same chromosome number (74), but there are major differences between wild vicuna and guanaco in the growth of their incisor teeth, and in their behavior. The long-held view that both the llama and alpaca are descended from the wild guanaco has recently been challenged by the suggestion that the fine-wooled alpaca is the product of cross-breeding between the vicuna and llama (or guanaco). Based upon wool length and body size, there are two widely recognized breeds each for the llama and alpaca. The *suri* alpaca is well known for both long-straight and wavy wool fibers. Today no llamas or alpacas live independently of man in their native Andean homeland.

The center of lamoid domestication may have been the Lake Titicaca pastoral region or perhaps the Junin Plateau, 62mi (100km) to the northwest, where South American camelid domestication has been dated at 4,000–5,000 years ago.

One school believes that the dromedary was first domesticated in central or southern Arabia before 2000BC, others as early as 4000BC, while some place it between the 13th and 12th centuries BC. From there, domesticated camels spread to Egypt and North Africa, and later to East Africa and India.

The Bactrian camel was domesticated independently of the dromedary, probably before 2500BC at one or more centers in the plateaux of northern Iran and southwestern Turkestan. From here they spread east to Iraq, India and China.

The South American cameloids are seasonal breeders and mate lying down on their chests—copulation lasting 10–20 minutes. They are induced ovulators, give birth to only a single offspring in a standing position and neither lick the newborn nor eat the afterbirth. The newborn is mobile and following its mother within 15–30 minutes. The female comes into heat again 24 hours after birth, but usually does not mate for two weeks. Large-bodied llamas that have access to high levels of nutrition may breed before one year old, but most female lamoids first breed as two-year-olds.

There is no information on the social organization of the domesticated llama and alpaca, because non-breeding males are castrated. However, circumstantial evidence suggests a territorial system, with

▲ **The proud, aloof face** of the dromedary. Actually, the facial characteristics which in human terms imply arrogance and disdain are meaningless in camelid terms. A camel with ears pressed flat to the body is far more likely to be hostile than one apparently sneering. Dromedary camels in Africa are generally maintained in a semi-wild state in which they are free to forage alone but dependent on man for water. During the rutting season they are guarded by their owners. Unguarded camels form stable groups composed of up to 30 animals. Free-grazing camels in North Africa form bachelor herds and mixed herds of perhaps one male plus 10–15 females and their young. Unlike the South American lamoids, male camels possess an occipital or poll gland immediately behind the head, which plays a role in the camel's mating behavior. The male camel also occasionally inflates his oral skin bladder ("dulaa") as a part of his rutting display towards other males.

males maintaining a harem of breeding females (polygynous).

The vicuna is strictly a grazer in alpine *puna* grassland habitats between 12,000–15,700ft (3,700–4,800m). They are sedentary, non-migratory with year-round defended feeding and sleeping territories. Populations are divided into male groups and family groups. The territories are defended by the territorial male and are occupied by the male, several females and their offspring of less than one year old. The territory is in two parts—a feeding territory where the group spend most of the day, and a sleeping territory located in higher terrain; the corridor between the two is not defended. The feeding territory—average size 45 acres (18.4ha)—is a predictable source of food; is the place where mating and birth often occur; and is a socially stable site with the advantages of group living, where females can raise their offspring. Sleeping territories average 6.4 acres (2.6ha).

The guanaco, which is twice as large as the vicuna, is both a grazer and browser, may occur in either sedentary or migratory populations, is much more flexible in its habitat requirements, in that it occupies desert grasslands, savannas, shrublands, occasionally forest, and ranges in elevation between 0–13,900ft (0–4,250m). Unlike the vicuna, it does not need to drink. Its territorial system is similar to that of the llama and alpaca, but the number of animals using the feeding territory, unlike the vicuna, is not related to forage production. Vicuna and guanaco young are forcefully expelled from the family group by the adult male, as old juveniles in the vicuna, and as yearlings in the guanaco.

Camels mate throughout the year, but a peak of births coincides with the season of plant growth. Copulation again takes place lying down, and they may be induced ovulators. Litter size is one. Camels eat a wide variety of plants over expansive home ranges. They eat thorns, dry vegetation, and saltbush that other mammals avoid. Camels can endure long periods without water (up to 10 months if not working), and so can graze far from oases. When they do drink, they can consume 30 gallons (136 liters) within a short time. They conserve water by producing dry feces and little urine, and allow their day-time body temperature to rise by as much as 11–14.5°F (6–8°C) during hot weather to diminish the need for evaporative cooling by sweating, although sweat glands occur over most of their body for use when necessary. Their nostrils, which can be closed to keep out blowing sand, their nostril cavities, which reduce water loss by moistening inhaled air and cooling exhaled air, and the localized storage of energy-rich fats in the hump, help them to survive long periods without food. A thick fur and underwool provide warmth during cold desert nights and some daytime insulation against the heat.

Today there are approximately 21.5 million camelids. In South America there are an estimated 7.7 million lamoids, with 53 percent in Peru, 37 percent in Bolivia, 8 percent in Argentina, and 2 percent in Chile. The domestic llamas and alpacas (91 percent of the total) are far more numerous than the wild guanacos and vicunas (9 percent). Llamas (3.7 million) are slightly more abundant than alpacas (3.3 million), and guanacos (575,000) are much more common than vicunas (85,000). Most alpacas (91 percent) and vicunas (72 percent) are in Peru, the majority of South American's llamas (70 percent) are in Bolivia, and nearly all the guanacos (96 percent) are found in Argentina. In general, numbers of alpaca and vicuna are increasing due to the value of their wool, whereas llamas are declining as they are replaced by trucks and rail transport. The guanaco is decreasing, due to hunting for its wool and skin, and competition with livestock.

Ninety percent of the world's 14 million camels are dromedaries, with 63 percent of all camels in Africa. Sudan (2.8 million), Somalia (2.0 million), India (1.2 million),

and Ethiopia (0.9 million) have the largest populations of camels, with Somalia having the highest density (8 per sq mile) by far. Numbers of camels have drastically declined over recent decades in some countries (eg Turkey, Iran and Syria), due in part to the forced settlement of nomads, but the worldwide camel population has remained relatively stable. A large but uncounted population of feral camels occupies the arid

▲ **The curiously craning gait** of vicuna, the smallest of the South American cameloids. Reduced to dangerous population levels in the 1960s, they are now recovering under protection.

▶ **Guanaco on the pampas.** Millions of these elegant camelids once roamed the South American plains, but their numbers are much reduced and they are still hunted in Argentina.

Abbreviations: SH = shoulder height.   wt = weight.   Approximate nonmetric equivalents 2.5cm = 1in; 1kg = 2.2lb.

D Domesticated.   V Vulnerable.   * CITES listed.

Llama D
Lama glama

Andes of C Peru, W Bolivia, NE Chile, NW Argentina. Alpine grassland and shrubland 7,500–13,000ft. HBL 47–88in; SH 43–47in; wt 285–340lb.
Coat: uniform or multicolored white, brown, gray to black. Gestation: 348–368 days. Two breeds: *chaku, ccara*.

Alpaca D
Lama pacos

Andes of C Peru to W Bolivia. Alpine grassland, meadows and marshes at 14,500–15,750. HBL 47–88in; SH 37–41in; wt 121–145lb.

Coat: uniform or multicolored white, brown, gray to black; hair longer than llama. Gestation: 342–345 days. Two breeds: *huacaya, suri*

Guanaco *
Lama guanicõe

Andean foothills of Peru, Chile, Argentina, and Patagonia. Desert grassland, savanna, shrubland and occasional forest at 0–14,000ft. SH 43–45in; wt 220–265lb.

Coat: uniform cinnamon brown with white undersides; head gray to black. Gestation: 345–360 days. Four questionable subspecies.

Vicuna V
Vicugna vicugna

High Andes of C Peru, W Bolivia, NE Chile, NW Argentina. Alpine *puna* grassland at 12,100–15,750ft. SH 34–38in; wt 99–120lb.
Coat: uniform rich cinnamon with or without long white chest bib; undersides white. Gestation: 330–350 days. Two subspecies: Peruvian, Argentinean.

Dromedary D
Camelus dromedarius
Dromedary, Arabian camel or One-humped camel.

SW Asia and N Africa. Deserts. Feral

in Australia. Height at hump 75–90in; wt 990–1,450lb. Coat color variable from white to medium brown, sometimes skewbald; short but longer on crown, neck, throat, rump and tail tip. Gestation: 390–410 days.

Bactrian camel V
Camelus bactrianus
Bactrian or Two-humped camel.

Mongolia. Steppe grassland. Height at hump 75–90in; wt 990–1,450lb. Coat: uniform light to dark brown; short in summer with thin manes on chin, shoulders, hindlegs and humps; winter coat longer, thicker and darker in color. Gestation: 390–410 days.

and train transport. Llamas can carry loads of 55–132lb (25–60kg) for 9–18mi (15–30km) a day across rugged terrain. Llamas have been introduced into a number of countries in small numbers for recreational backpacking, wool production, and novelty pets. In the Andes, the woolly alpaca is replacing the llama as the most important domestic lamoid.

There were millions of guanacos in South America when the Spaniards arrived, perhaps as many as 35–50 million on the Patagonian pampas alone. Today, the guanaco is still the most widely distributed lamoid, but in greatly reduced numbers compared to historical times. The guanaco is protected in Chile and Peru, but not in Argentina, where tens of thousands of adult and juvenile (*chulengos*) guanaco pelts are exported annually.

From a population level of millions in the 1500s, to 400,000 in the early 1950s, then to less than 15,000 in the late 1960s, the vicuna was placed on the rare and endangered list by the International Union for the Conservation of Nature (IUCN) in 1969. Today the vicuna is fully protected and populations are increasing. World vicuna population size has recovered to over 80,000 and its status was changed by IUCN from endangered to vulnerable in 1981.

Dromedary camels are important beasts of burden throughout their range, but they are especially valued for their meat, wool, and milk—1.3 gallons (6 liters) per day for 9–18 months. Man in return provides water for the camels. The milk produced by camels is often the main source of nourishment for the desert nomads of the western Sahara. Nomads use camels as work animals primarily for moving camp and carrying water. They walk 19mi (30km) a day at a leisurely pace to allow for feeding and resting, and carry up to 220lb (100kg) per animal. Camels play a special role in social customs and rituals. They are required as gifts for marriage and reparation for injury and murder. White camels are favored most. Camels have been used to transport military material and personnel during conquests and wars of North Africa, beginning with Napoleon's campaign there in 1798.

The future of the camel in the Saharan countries depends upon the fate of the nomads. If these countries settle their nomads, the vast stretches of arid lands will be of little use. If the nomads are encouraged to continue their traditional ways, the desert will continue to serve as an important resource for millions of people and camels.

inland regions of the mainland of Australia.

The once vast range of the Bactrian camel has contracted severely in recent times, although some remain in Afghanistan, Iran, Turkey, and the USSR. Most of the camels in Mongolia and China are Bactrians, and there are probably less than 1,000 wild Bactrians in the trans-Altai Gobi Desert. These animals have slender build, short brown hair, short ears, small conical humps, small feet, no chest callosities, and no leg callosities.

The camelids' energy as pack animals and their products of meat, wool, hair, milk and fuel have been indispensable to man's successful habitation and survival in earth's temperate and high-altitude deserts. The Inca Empire's culture and economy revolved around the llama, which provided the primary means of transportation and carrying goods. Domestic cameloid numbers declined drastically during the century following the Spanish invasion of the Central Andes. Uncontrolled shooting, disruption of the Inca social system, and the introduction of sheep were responsible for the crash.

Today, the docile llama is still actively used in the Andes as an important pack animal, though it is being replaced by trucks

CHEVROTAINS

Family: Tragulidae
Four species in 2 genera.
Order: Artiodactyla.
Distribution: tropical rain forests of Africa,
India and SE Asia.

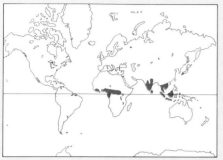

Water chevrotain ▣

Hyemoschus aquaticus
C and W Africa. Tropical rain forest; nocturnal.
Size: head-body length 28–32in (70–80cm);
shoulder height 13–16in (32–40cm); tail
length 4–5.5in (10–14cm); weight 18–29lb
(8–13kg).
Coat: blackish red-brown, with lighter spots,
arranged in rows and continuous side stripes;
throat and chest with white herringbone
pattern. Gestation: 6–9 months. Longevity:
10–14 years.

Spotted mouse deer

Tragulus meminna
Spotted mouse deer or Indian spotted chevrotain.
Sri Lanka and India. Tropical rain forest,
particularly rocky; nocturnal. Size: head-body
length 20–23in (50–58cm); shoulder height
10–12in (25–30cm); tail length 1.2in (3cm);
weight about 6.6lb (3kg). Coat: reddish brown
with lighter spots and stripes; more or less
similar to Water chevrotain. Gestation:
probably 5 months.

Lesser mouse deer

Tragulus javanicus
Lesser mouse deer or Lesser Malay chevrotain.
SE Asia. Tropical rain forest and mangroves;
nocturnal. Size: head-body length 17–19in
(44–48cm); shoulder height 8in (20cm); tail
length 2.5–3.2in (6.5–8cm); weight 4–5.7lb
(1.7–2.6kg). Coat: more or less uniform red with
characteristic herringbone pattern over the fore
part of the body. Gestation: probably 5 months.

Larger mouse deer

Tragulus napu
Larger mouse deer or Greater Malay chevrotain.
SE Asia excluding Java. Tropical rain forest;
mainly nocturnal. Size: head-body length
20–24in (50–60cm); shoulder height
12–14in (30–35cm); tail length 2.8–3.2in
(7–8cm); weight 9–13lb (4–6kg). Coat:
more or less uniform red. Gestation: 5 months.

▣ CITES listed.

THE smallest species no bigger than a rabbit, virtually unchanged in 30 million years of evolution and intermediate in form between pigs and deer—these are the main claims to distinction of chevrotains, diminutive inhabitants of Old World tropical forests. During the Oligocene and Miocene (38–7 million years ago), chevrotains had a worldwide distribution, but today the four species are restricted to the jungles of Africa and Southeast Asia.

Chevrotains are among the smallest of ruminants—hence the collective name mouse deer—with the Lesser mouse deer the smallest at a mere 4.4lb (2kg). All species have a cumbersome build, with short, thin legs and limited agility. Males are generally smaller than females: in the Water chevrotain, 20 percent less in weight. The coat is usually a shade of brown or reddish brown, variously striped or spotted depending on the species. They are shy, secretive creatures, mostly active at night. They inhabit prime tropical forest, the three Asiatic species showing a preference for rocky habitats. The Water chevrotain is a good swimmer. Their diet mainly comprises fallen fruit, with some foliage.

Chevrotains are ruminants, ie their gut is modifed for the fermentation of their food, but they exhibit many non-ruminant characters, and should be regarded as the most primitive ruminants, providing a living link between non-ruminants and ruminants. They have a four-chambered stomach, although the third chamber is poorly developed. Other anatomical features chevrotains share with the ruminants are: a lack of upper incisor teeth; incisor-like lower canines adjoining a full set of upper incisors; and only three premolar teeth. Behavioral similarities with ruminants include: a typical (although extreme) solitary life-style for a forest ruminant; single young; and inges-

tion of the placenta by females after birth. More interestingly, chevrotains share a number of (primitive) characteristics with non-ruminants. These include: no horns or antlers; continually growing projecting upper canines in males (peg-like in females); premolars with sharp crowns; and all four toes fully developed. Behavioral patterns shared with pigs include: long copulation and simple sexual behavior; lack of visual displays, so communication is limited to smells and cries; lack of specialized scent glands below the eyes or between the toes; and a habit of lying down rump first, then retracting their forelimbs beneath them. Of the four species, the Water chevrotain is the most pig-like (ie primitive) since its forelimbs lack a cannon bone and the skin on the rump forms a tough shield which protects the animal against canine teeth.

The reproductive biology of chevrotains is poorly known. In the Water chevrotain and Larger mouse deer, breeding occurs throughout the year, with only one young per litter; weaning occurs at 3 months in both species with sexual maturity achieved at 10 months and 4–5 months, respectively. In the former species, young are known to be hidden in undergrowth soon after birth, but maternal care is limited to nursing. In the Larger and Spotted mouse deer, mating occurs within two days of birth. In the Water chevrotain, the only mating "display" is a cry by the male akin to the *chant de cour* of pigs which brings the courted female to a standstill for copulation, which may last 2–5 minutes. Water chevrotains are solitary except during the mating season, and mostly show no interest, aggressive or otherwise, in other members of their species. Communication is by calls and scent.

Water chevrotains possess anal and preputial glands, and this species marks its home range with urine and feces, the latter

▲ **The Lesser mouse deer,** the smallest of the chevrotains.

◄ **The Spotted mouse deer** ABOVE, a small primitive artiodactyl intermediate in form between pigs and deer.

impregnated with anal gland secretions. All species have a chin gland which is rudimentary in the Water chevrotain and Spotted mouse deer. In males of the other species it produces copious secretions which they use to mark a mate's or male antagonists's back during encounters, In the Water chevrotain, the heaviest and oldest animals are dominant. Owing to their solitary life, however, there is no established hierarchy. Fighting between males is reduced to a short rush, each antagonist biting his opponent all over his body with his sharp canines.

Spacing behavior is only known for Water chevrotains. They mainly inhabit terrestrial habitats, but will retreat to water when danger threatens. Home ranges are 60–70 acres (23–28ha) in males and 32–35 acres (13–14ha) for females, and they always border a watercourse at some point. Home ranges of adult females do not overlap: neither do those of males, although they do overlap those of females. There is no evidence of territorial defense. Population density varies from 20–72 per sq mi (7.7–28 per sq km) for the Water chevrotain in Gabon and 1.5 per sq mi (0.58 per sq km) for the Spotted mouse deer in Sri Lanka.

Their small size makes chevrotains easy and important prey for various predators, such as big snakes, crocodiles, eagles and forest-inhabiting cats. As with many other tropical forest species, the survival of these four living fossils depends on conservation of their habitat and restriction of hunting.

GD

MUSK DEER

Family: Moschidae ✳
Three species of the genus *Moschus*.
Order: Artiodactyla.
Distribution: Asia.

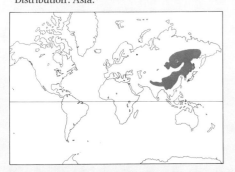

Habitat: mountain forests; active in twilight.

Size: head-body length 31–39in
(80–100cm); tail 1.5–2.5in
(4–6cm); height 20–28in
(50–70cm); weight 15.5–37.5lb
(7–17kg).

Musk deer

*Moschus moschiferus, M. berezovskii,
M. chrysogaster*†
Distribution: Siberia, E Asia, Himalayas.
Coat: grayish-brown to golden, speckled;
striped at the under part of the neck.

Gestation: 150–180 days.

Longevity: 13 years (in captivity).

†As there is very little information on the
distribution, biology and morphology of the
three species, which until recently have been
considered a single one with many subspecies,
the above data refer to the genus as a whole.

✳ CITES listed.

► **The "milk-step."** Suckling in Musk deer
is unusual in that the young touches the
mother's hindleg with a raised foreleg. This
gesture is similar to that seen during the
courting behavior of other hoofed mammals,
such as the gerenuk. The female usually gives
birth to one young, rarely twins. The newborn
calves are very small in comparison with their
mother—20–25oz (600–700g). During the
first weeks of life they are motionless and
inconspicuous among rocks or in dense
undergrowth. From time to time, the mother
visits the young and lets it suckle; after a month
it leaves the resting place and accompanies the
mother.

Musk, a substance produced by a gland in
the male Musk deer, is an important
and expensive ingredient of the best perfumes. In 1972 an ounce of musk was more
valuable in Nepal than an ounce of gold!
Despite its economic importance and its
wide distribution, little is known of the
biology of the Musk deer. This is especially
regrettable, since in many regions it is
declining due to hunting or deforestation.

Musk deer are stockily built animals with
small heads. The hindlegs are about 2in
(5cm) longer than the forelegs, indicating a
tendency to move by leaping. The hooves,
including the lateral toes, are long and
slender. Most of the hair is very coarse.
Neither sex possesses antlers but the male
has long, saber-like upper canine teeth
which project well below the lips; in females
the canines are small and not visible. In this
respect, Musk deer resemble other primitive
deer like the Water-deer. They give a good
impression of the appearance of the earliest
predecessors of antler-bearing deer. Their
most notable, and unique, feature is the
musk bag or pod which develops when the
male has reached sexual maturity. This sac,
situated between the genitals and the umbilicus, is about the size of a clenched fist. Its
glands produce a secretion whose precise
function is unknown but which is probably
a signal for the females. In contrast to the
adults, newborn Musk deer have a distinctly
striped and spotted coat. Musk deer are shy
and furtive animals with a keen sense of
hearing.

The classification of the Musk deer has
long been confused. Formerly, only one
species was recognized, but recently the
existence of three different species has been
proven. The limits of their distribution are
unknown, so the map gives only an idea of
the distribution of the genus as a whole.

Musk deer, being primarily forest
animals, are never found in desert regions or
in areas with a dense human population.
Musk deer prefer dense vegetation, especially hills with rocky outcrops in coniferous,
mixed or deciduous forests. In the morning
and evening hours they leave their resting
places, situated among rocks or fallen trees,
in search of food. Their diet consists of
leaves, flowers, young shoots and grasses,
also of twigs, mosses and lichens, especially
in winter. In captivity, they readily accept
lettuce, carrots, potatoes, apples, rolled oats,
hay and alfalfa.

Musk deer are solitary for most of the
year. Outside the rutting season, more than
two or three are seldom seen together; such
groups usually consist of a female and her
young. During the rutting season, mainly
November/December in northern latitudes,
males run restlessly, pursuing females,

chasing them to the point of exhaustion, when they seek shelter. During this period, males fight for the females, and their teeth can inflict deep, and sometimes deadly, wounds on their opponent's neck and back. During the weeks of courtship, the males eat little, they are very excited and cover large distances. After the rutting period they return to their original territories.

Musk deer are strongly territorial. Within their home range they regularly use well-established trails which connect feeding places, hiding places and the latrines where their droppings are deposited. By continuous use, these spots may reach considerable size. Both sexes cover their droppings immediately after defecation by scratching the soil with their forelegs. Males scent mark tree trunks, twigs and stones within their territories by rubbing their tail gland against them, producing oily patches. When disturbed, they dart off in enormous bounds. Thanks to the strongly developed lateral toes, Musk deer climb well on precipitous crags, rocks and even trees. They also run readily on snow or heavy soil. Apart from man, they are subject to predation by lynx, wolf, marten, fox and probably some birds of prey.

From time immemorial, Musk deer have been hunted by man for musk, one of the few mammal substances used in perfumes. Musk is not only used in Europe as a base for exquisite perfumes but also in the Far East for a great variety of medicinal applications (sore throats, chills, fever and rheumatism). In a single year, Japan, for instance, imported 11,000lb (5,000kg) of musk which, at an ounce per pod, represents the astonishing total of 176,000 male Musk deer. As the animals are taken mainly with automatic snares set up on their trails, the total loss, including muskless females and young, must have been even higher. Because of the relentless persecution, Musk deer have already become rare in parts of their range.

It is, however, not necessary to kill the males in order to obtain the musk. Therefore, in 1958, the Chinese started a breeding program in the province of Sichuan. Between 1958 and 1965, 1,000 Musk deer were caught alive. In spite of heavy losses at the beginning, mainly during the transportation and acclimatization period, the Chinese have succeeded in breeding the species in vast enclosures. Although juvenile mortality is still high and longevity relatively short (the same applies to zoos, where Musk deer are seldom on exhibit), some satisfactory results have been achieved. When the sexual activity of the male is at its peak, the animals are caught by hand. The musk is then removed from the pod by a spoon inserted into the aperture of the sac, a procedure which takes only minutes; then the animal is released. The musk, a jelly-like, oily substance with a strong odor and reddish-brown color is subsequently dried, when it becomes a powdery mass which gradually turns black.

If the techniques of breeding and handling Musk deer continue to improve, their farming may help to reduce the pressure on wild populations and may make a valuable contribution to the conservation of this interesting little deer.   HF

▼ **The long, protruding canines** of the male Musk deer give it a lugubrious expression. The Musk deer is thought to be similar to the ancestors of the antler-bearing deer.

DEER

Family: Cervidae
Thirty-six species in 16 genera.
Order: Artiodactyla.
Distribution: N and S America, Eurasia,
NW Africa (introduced to Australasia).

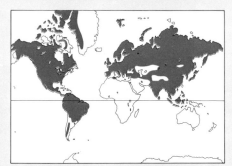

Habitat: mainly forest and woodland, but also
arctic tundra, grassland and mountain regions.

Size: shoulder height and weight from 15in
(38cm) and 17.5lb (8kg) in Southern pudu to
90in (230cm) and 175lb (800kg) in the
moose.
Coat: mostly shades of gray, brown, red and
yellow; some adults and many young spotted.

Gestation: from 24 weeks in Water-deer to 40
weeks in Père David's deer.

Species include: **Water-deer** (*Hydropotes
inermis*); **muntjacs** (5 species of genus
Muntiacus); **Tufted deer** (*Dama dama*); **chital**
(*Axis axis*); **Swamp deer** (*Cervus duvauceli*); **Red
deer** (*Cervus elaphus*); **wapiti** (*Cervus
canadensis*); **Sika deer** (*Cervus nippon*); **Père
David's deer** (*Elaphurus davidiensis*); **Mule deer**
(*Odocoileus hemionus*); **White-tailed deer**
(*Odocoileus virginianus*); **Roe deer** (*Capreolus
capreolus*); **moose** (*Alces alces*); **reindeer**
(*Rangifer tarandus*); **Pampas deer** (*Ozotoceros
bezoarticus*); **huemuls** (2 species of genus
Hippocamelus); **Southern pudu** (*Pudu pudu*);
Northern pudu (*Pudu mephistophiles*); **Red
brocket** (*Mazama americana*); **Brown brocket**
(*Mazama gouazoubira*).

T HE commonest view of a wild deer is of a
rapidly disappearing rump, but even
that is rare because deer are always on
guard against predators. Yet cave paintings
dating from 14,000 years ago and more
modern images, such as The Monarch of the
Glen, Bambi and Santa Claus' reindeer,
show the interest which deer hold for man.
Firstly deer were a source of food and then
sport, to the extent that the Norman kings of
England planted forests in the South of
England to ensure a supply of deer for the
chase. More recently deer have been regarded as creatures of interest in their own
right and knowledge of their lives and
behavior has broadened.

Deer are similar in appearance to other
ruminants, particularly antelopes, with
graceful, elongated bodies, slender legs and
necks, short tails and angular heads. They
have large, round eyes which are placed
well to the side of the head, and triangular or
ovoid ears which are set high on the head.
The size of deer ranges from the moose to the
Southern pudu which weighs only about 1
percent of the moose's 1,750lb (800kg).

The antlers of adult male deer distinguish
them from other ruminants. These are bony
structures like horns, but unlike horns they
are shed and regrown each year. In form
they range from single spikes with no
branches, as in pudu, to the complex,
branched structures of Red deer, Fallow
deer and reindeer (or caribou). These last
two species have palmate antlers where
the angle between some of the branches is
filled in, giving a flat "palm" with the points
of the branches resembling fingers from the
palm. Only one species—the Water-deer—
lacks antlers, but the males have tusks, as do
muntjacs and Tufted deer which have only
small simple antlers.

Females do not have antlers except in the
reindeer. Many explanations have been suggested for this; perhaps the most likely is

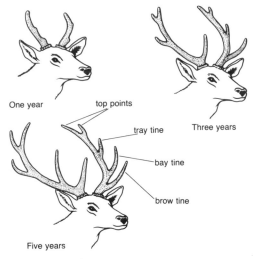

One year   top points   Three years
tray tine
bay tine
brow tine
Five years

that they help females compete for food in
the large, mixed sex herds, unusual for deer,
in late winter when the reindeer dig craters
in the snow to reach the vegetation below.

Why do stags make such a huge investment in antler growth each year? Antlers
are used as weapons in fights for females and
are often damaged. In Red deer major
damage to the antlers can reduce rutting
success that season, and it would have the
same effect in future years if antlers were not
regrown each year. The effect of antlers on
dominance is particularly clear at casting
time in Red deer when the older, dominant
stags who cast first can no longer displace
younger ones with antlers.

The color of the coat varies through many
shades of gray, brown, red and yellow.
Usually the underparts are lighter than the
back and flanks, and there is often a light
colored area around the anus—the rump
patch—which may be fringed with dark
hair. The rump patch is often made conspicuous by alarmed deer, who raise their
tails and rump hair as they flee. The young
of many species have a pattern of light
colored spots on a darker ground which
improves their camouflage; this pattern is
retained by adults in some species, eg Sika

◄ **Antler development in Red deer.** Deer are unique among ruminants in their annual cycle of antler growth and shedding. The process is under the control of the sexual and growth hormones. Fur-covered skin (velvet) carries blood to the growing antlers. In the fall the velvet dries up and the deer thrash their antlers against vegetation to remove it. In the winter the antlers are shed. Young deer take several years to develop a full set of antlers, with bifurcated points and projecting fines.

◄ **A forest of antlers.** This herd of Red deer stags on a deer farm seem almost part of the vegetation. Herds of stags have been kept for their velvet, an expensive ingredient of oriental medicines, and sometimes for their meat.

▼ **Gory antlers** of a reindeer bull which has just shed its velvet.

and Fallow deer. The coat is molted twice a year, in spring and fall. Deer are more or less completely covered in hair, but all species except the reindeer have a small naked patch on the muzzle.

In common with other ruminants, deer bite off their food between the lower incisors and a callous pad on the upper gum. The dental formula of adult deer is I0/3, C0/1, or C1/1, P3/3, M3/3. The presence of upper canines depends on the species, and in male Water-deer, muntjacs and Tufted deer they develop into tusks which may be used for fighting. The premolars and molars grind food during rumination (chewing the cud).

Deer use hearing, smell and sight to detect danger from predators. When feeding, the first warning a deer has of a predator is probably through sound or smell, as sight is restricted by the vegetation. Deer, however, raise their heads and scan their environment during feeding which probably increases their rate of predator detection. When a deer is alarmed it will raise its head high, stare hard at the source of alarm and rotate its ears forwards towards it. Although deer are quick to see and respond to movement, they may fail to detect, for example, a human stalker if he keeps quite still until the deer begins to feed again. If they hear or see movement deer rapidly flee.

Deer also use smell to gain information about members of their own species. A rutting Red deer stag detects which hinds in his harem are in heat by sniffing them or the earth where they have lain or urinated. Most species have interdigital (or foot) glands which probably leave scent trails, and all have facial (or pre-orbital) glands which produce strong-smelling secretions.

Vocal communication among females and young deer is more or less limited to bleats by which mothers and offspring locate each other and alarm barks which are given when they are disturbed. However, a wide range of vocalizations, including the whistling scream of Sika deer, barks of muntjac and the bellowing roar of Red deer, is produced by males competing for females in the breeding season.

Deer are typically woodland animals, not found far from cover in open country; the largest number of species occur in subtropical woodland. However, deer occur in almost all habitats from arctic tundra (reindeer) to open grassland in Argentina (Pampas deer) and deep forest (pudu). Deer are indigenous to Europe, Asia, North, Central and South America and northern Africa. However, they have been introduced into Australia, New Zealand and New Guinea. There have also been translocations of species within the natural range of deer, for example England has feral muntjac originating from India, and there are feral Red deer and Fallow deer originating from Europe in South America.

Most species of deer can be distinguished on the basis of size, coat color or other external characteristics such as tail length and ear size. Some species are extremely easily recognizable—nobody could mistake a huge, ungainly deer with humped shoulders, a belled throat and palmate antlers for anything but a moose. In other cases the differences between species are more subtle, for example the Red and Brown brockets differ only in a slight variation in the shade of coat color. Size and form of the antlers are also distinctive: for example the dichotomously branched antler of an adult male Mule deer is quite different from that of the adult male White-tailed deer where tines branch off a main beam. But not all members of a species are adult males with antlers, and there may be confusion between the antlers of immature males of Mule and White-tailed deer; here a minor feature of the antler—the length of the basal snag (a small point)—differs between the species. However, this still leaves the problem of identification of deer without antlers, males after they have cast their antlers and females who never have them. In these circumstances tail length and the length of the metatarsal gland are used to distinguish Mule deer and White-tailed deer.

In some cases where two species are very similar but occupy distinct ranges, it is more convenient to separate them; for example, the two species of huemul that live in widely different parts of the Andes. In other instances no such geographical separation exists between very similar species. For example, Red deer are found from western Europe to central Asia and wapiti from central Asia to North America. Both show considerable variation between subspecies and localities but are basically similar to each other. On this basis some authorities prefer to classify them as the same species. However, since not only the coloration but also the vocalization of rutting males are different, they are treated here as separate species.

Some authorities regard the three species of musk deer (genus *Moschus*) as members of this family. However, others place them in a separate family, as here (see pp78–79).

Deer generally feed by grazing on grass or browsing on the shoots, twigs, leaves, flowers and fruit of herbs, shrubs and trees. However, chital and Swamp deer in India feed almost exclusively on grass (but see box) while the Indian muntjac prefers herbs, leaves, fruit, mushrooms and bark. Reindeer

and caribou scrape away snow in winter to get at lichens, which form a major part of their diet.

Within a species the type and amount of food eaten is determined by the individual's nutritional requirements and by food availability. Large animals such as deer have higher maintenance requirements than herbivores such as rodents, so in the many species where males are larger than females, males might be expected to need more food. However, deer need food for reproduction as well as for maintenance and comparisons of male and female nutritional requirements are complicated by differences in the extent and timing of the costs of reproduction. Consider Red deer: the highest requirements for energy and protein by hinds occur in spring and early summer during the last third of pregnancy and early lactation; the peak for stags is slightly later when they grow antlers and neck muscle in the approach to the rut. In a study of wild Scottish Red deer, stags spent more time in summer feeding than hinds without calves (10.4 and 9.8 hours per day), reflecting the high cost of milk production. In winter, stags increased the amount of time spent feeding to 12.88 hours per day; hinds with calves still fed for 11.08 hours per day even though most of the calves were weaned, probably because it took longer to ingest the same amount of food as the available vegetation declined.

Food availability is low in winter for many northern deer and they use more energy than they gain, thus losing weight. That reduced food intake is due to a reduction in appetite as well as in food availability was shown in captive Red and White-tailed deer which reduced intake in winter even when given unlimited food. Presumably in the natural state this helps prevent them from wasting more energy searching for low-quality food than the food would provide.

Another response to low food availability is to switch to a different diet. Thus the

▲ **The noble features** of the Rusa deer, an Indonesian species that has been introduced to Australia, New Zealand, Fiji and New Guinea.

▼ **Representative species of deer.** (**1**) Southern pudu (*Pudu pudu*). (**2**) Pampas deer (*Ozotoceros bezoarticus*). (**3**) Marsh deer (*Blastocerus dichotomous*). (**4**) Peruvian huemul (*Hippocamelus antisensis*). (**5**) Red brocket (*Mazama americana*). (**6**) White-tailed deer (*Odocoileus virginianus*). (**7**) Reindeer (*Rangifer tarandus*).

ing lifetime reproductive success are different in males and females. In males, reproductive success depends on the number of receptive females to which a male has access. Females are highly unlikely to lack access to males and their reproductive success depends on their ability to rear offspring and the factors that affect that ability. A study of wild Red deer on the Isle of Rhum has attempted to tease out some of these factors. Estimates of lifetime reproductive success of stags ranged from 0–24 calves with an average of 4.2. Lifetime reproductive success depends on lifespan and the success a stag has within breeding seasons. Success within a season was related to harem size, the duration of harem holding and the timing of holding relative to the conception peak; the principal determinant of these aspects of rutting success was the ability to beat other males in fights. This in turn depended on body size and condition, and indeed lifetime reproductive success of stags was correlated with antler weight, which is an indicator of body size. As a rule, large stags leave more offspring than small ones.

The total number of offspring reared by females varied from 0–13 with an average of 4.5. The lifetime reproductive success of a hind is affected by her lifespan, the number of years in which she fails to conceive (usually because of poor condition) and the mortality among her calves. In fact fecundity was less important than calf mortality.

12   13   14   15

which was in turn related to birth date, late calves being more likely to die in their first winter. Conception date and therefore birth date were related to the mother's condition during the rut. The size and condition of calves was also related to the mother's condition, but not to her skeletal size. Thus mothers with best access to the food needed to maintain good body condition successfully rear the most calves.

Because body size is important to male reproductive success and early differences persist into adulthood, natural selection favors rapid growth in males. It also favors mothers who invest heavily in their sons' early growth since they will leave more grandchildren if their sons are large. There is evidence from the longer gestation length of male calves (236 days rather than 234 for females) and the higher proportion of mothers of male calves who are in such poor condition that they fail to conceive in the following rut that mothers do make a heavier early investment in sons than in daughters.

Skeletal size and early growth have less effect on hind reproductive success than access to food during reproduction. The importance of access to food may explain why hinds remain in their natal areas when adult and why mothers allow them to share their resources (see below). Such tolerance of adult daughters may balance the heavier early investment in sons.

The social organization of different species is probably related to their differing food supply. The food of browsing deer such as Roe deer or the brockets occurs in small, discrete patches, so close proximity of other animals is likely to interfere with feeding; these species live singly or in small groups. Deer which graze in more open habitats live in larger groups (eg wapiti, Sika deer); feeding interference is not as important for grazers and they may gain some protection from predators by their companions' vigilance. In most deer, males and females live apart outside the breeding season and the social organization of the two sexes may be very different, as a consideration of Red deer will illustrate.

Males live in bachelor groups whose membership varies from day to day but in which there are consistent, linear dominance hierarchies related to body size. Threats often displace the victim from a particular feeding site and are more frequent in winter when food is scarce than in summer (1.8 versus 1.0 threats per stag per hour). High ranking stags are thus able to monopolize good food patches. Stags do not

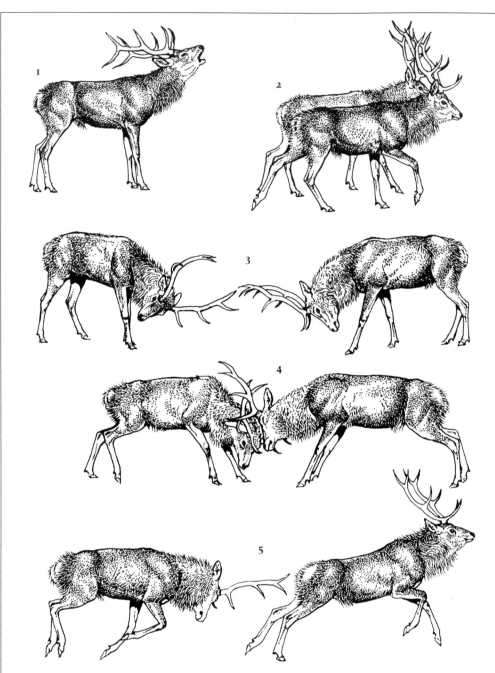

The Red Deer Rut

Rutting Red deer stags fight vigorously for the possession of harems. Usually a challenging stag approaches a harem holder and the two roar at each other for several minutes (1). Then the contest progresses to a parallel walk (2), the stags moving tensely up and down until one of them turns and faces his opponent, lowering his antlers (3). The other stag turns quickly, they lock antlers and push strenuously against each other, each trying to twist its opponent round and to gain the advantage of slope (4). When one succeeds in pushing the other rapidly backwards the loser disengages and runs off (5). Many contests do not reach the fight stage, but end after the roaring contest or the parallel walk with the challenging stag withdrawing.

A mature stag will fight about five times over the period of the rut, risking considerable danger and cost. In a study on the Isle of Rhum, Scotland, about 23 percent of mature stags were injured during the rut each year, up to 6 percent permanently. Even if he won the fight the harem holder often lost hinds to the young stags who lurk on the edges of harems waiting for the opportunity to steal them. With such costs it is not surprising that stags are reluctant to fight. Unequal contests between stags of widely differing fighting success were less common than well-matched ones, suggesting that stags were able to assess their chances of winning.

Playback experiments showed that stags attempted to outdo their rivals in roaring frequency, and the roaring rates of individual stags were correlated with fighting success.

▲ **Wapiti in the snow.** In the Rocky Mountains of the USA and Canada, wapitis have been forced by farming into more inhospitable habitats. Many can die during harsh winters.

appear to distinguish relatives from non-relatives.

In contrast, hinds associate with relatives and threaten them less than non-relatives, and this is probably why threats are less frequent among hinds than among stags (0.37 threats per hind per hour in winter, 0.15 in summer). This sex difference in the treatment of relatives may occur because of kin selection, since a mother might reduce the number of grandchildren she leaves by denying her daughter access to food.

The two main factors which threaten deer populations are restriction and modification of habitat by human activity, and hunting, either as pests or for food, skins, antlers, etc. Nine species and several subspecies of deer are regarded as endangered.

Almost all species of deer have been hunted for meat, and many for their antlers as trophies. In parts of North America yearling wapiti males may be responsible for most of the following years' calves, as so many of the older males are shot for their antlers. The popularity of hunting has led to the introduction of deer to new ranges by man, and about eight species now exist in the wild in Australia and New Zealand, where deer are not indigenous.

Domestication of deer has until recently been restricted to the herding of reindeer by the Lapps for meat, hides and milk, and to the maintenance of deer (usually Red and Fallow deer) in parks for ornamental purposes. In the last two decades, however, there has been a major growth of Red deer farming in New Zealand, Australia, and Britain, chiefly for meat but also for the velvet covering the growing antlers which is believed by some, particularly in the Orient, to have aphrodisiac qualities.

Deer often come into conflict with man because they eat his crops, and they may also be pests of commercial forestry. Careful management is needed to ensure that the line between sensible pest control and persecution is not crossed, particularly in more vulnerable species.   RAC

THE 36 SPECIES OF DEER

Abbreviations: SH = shoulder height. wt = complete body weight. cwt = clean body weight, ie after removal of entrails. AL = length of antler beam. Pt = number of points on both antlers. ? = unknown. Dimensions are for males except where otherwise stated; females are usually smaller than males. Only males have antlers except in reindeer.

E Endangered. V Vulnerable. R Rare. I Indeterminate. * = CITES listed.

Subfamily Hydropotinae

Water-deer
Hydropotes inermis
Water-deer or Chinese water-deer

China, Korea. Swamps, reedbeds, grasslands. SH 20in male, 19in female; wt 24–31lb male, 18–24lb female. Antlers lacking in both sexes. Upper canines present and in males form tusks 3in long. Coat: reddish brown in summer, dull brown in winter; young dull brown with light spots each side of midline. Gestation: 176 days. Subspecies: 2.

Subfamily Muntiacinae

Indian muntjac
Muntiacus muntjac

India, Sri Lanka, Tibet, SW China, Burma, Thailand, Vietnam, Malaya, Sumatra, Java, Borneo (introduced in England). Woodland and forest with good undergrowth. SH 22in male, 19in female; wt 40lb. Antlers usually single spikes; antler pedicels extend down face; AL 7in. Pt 2–4. Upper canines present and in males form tusks 1in long. Coat: dark chestnut on back toning to almost white on belly, darker in winter than in summer. Gestation: about 180 days. Subspecies: 15.

Reeves's muntjac
Muntiacus reevesi
Reeves's or Chinese muntjac or Barking deer.

E China, Formosa (introduced in England). Habitat, antlers and canines as Indian muntjac. SH 16in; cwt 24lb. Coat: chestnut with darker limbs and black stripe along nape; rufous patch between pedicels on forehead; chin, throat and underside of tail white. Gestation: 210 days. Subspecies: 2.

Hairy-fronted muntjac I
Muntiacus crinifrons
Hairy-fronted or Black muntjac.

E China. Habitat, antlers and canines as Indian muntjac. SH 24in; wt? AL 2.5in, with small projection from inner side of base. Coat: blackish brown with lighter head and neck. Tail longer than in other muntjacs.

Fea's muntjac E
Muntiacus feae

Thailand, Tenasserim. Similar to Indian muntjac in all respects but coat darker brown with yellow hairs up to center of the pedicels.

Roosevelt's muntjac
Muntiacus rooseveltorum

N Vietnam. Intermediate in size between Indian and Reeves's muntjac. Coat: redder than Reeves's muntjac. In other respects same as Indian muntjac.

Tufted deer
Elaphodus cephalophus

S, SE and C China, NE Burma. Forest up to 15,000ft. SH 25in; wt? Antlers and pedicels smaller than in muntjacs; antlers unbranched and often completely hidden by tufts of hair which grow from forehead. Upper canines present and in male form tusks 1in long. Coat: deep chocolate brown, with gray neck and head; underparts, tips of ears and underside of tail white. Young have spots on back. Gestation: about 180 days. Subspecies: 3.

Subfamily Cervinae

Fallow deer *
Dama dama

Europe, Asia Minor, Iran (introduced to Australia, New Zealand). Woodland and woodland edge, scrub. SH 36in male, 31in female; wt 139–227lb male, 64–120lb female. Antlers branched and palmate; AL 28in. Coat: typically fawn with white spots on back and flanks in summer, grayish brown without spots in winter; rump patch white edged with black; black line down back and tail; belly white; color highly variable and melanistic (black), white and intermediate forms common. Young spotted in some color varieties. Gestation: 229–240 days. Subspecies: 2.

Chital
Axis axis
Chital, Axis deer or Spotted deer.

India, Sri Lanka (introduced to Australia). Forest edge, woodland. SH 36in; wt 190lb; AL 30in; Pt 6; beam of antler curves back and out in lyre shape. Coat: rufous fawn with white spots on back; no seasonal change in color. Young spotted. Gestation: 210–225 days. Subspecies: 2.

Hog deer
Axis porcinus

N India, Sri Lanka, Burma, Thailand, Vietnam (introduced to Australia). Grassland, paddyfields. SH 26–29in; wt 80–100lb; AL 15in; Pt 6. Coat:

yellowish brown with darker underparts, the metatarsal tufts lighter colored than the rest of the leg; young spotted. Build low and heavy with short face and legs. Gestation: about 180 days. Subspecies: 2.

Kuhl's deer R *
Axis kuhlii
Kuhl's or Bawean deer.

Bawean Island. SH 27in; wt? AL? Pt 6; pedicles long. Coat: uniform brown; tail long and bushy; young unspotted.

Calamian deer V
Axis calamianensis

Calamian Islands. Similar to Hog deer except face and ears shorter, and presence of white patch from lower jaw to the throat and white moustache mark.

Thorold's deer I
Cervus albirostris

Tibet. SH 48in; wt? Antlers branched; AL?; Pt 10. Coat: brown with creamy belly, white nose, lips, chin, throat, and white patch near ears; winter coat lighter in color than summer. Ears are narrow and lance shaped. Hooves are high, short and wide like those of cattle. Reversal of hair direction on withers gives the appearance of a hump.

Swamp deer E *
Cervus duvauceli
Swamp deer or barasingha.

N and C India, S Nepal. Swamps, grassy plains. SH 47–49in; wt 380–400lb. AL 35in; Pt 10–15. Coat: brown with yellowish underparts; in the hot season stags become reddish on the back. Young spotted. Subspecies: 2.

Red deer *
Cervus elaphus
Red deer, maral, hangul, shou, Bactrian deer or Yarkand deer.

Scandinavia, Europe, N Africa, Asia Minor, Tibet, Kashmir, Turkestan, Afghanistan (introduced to Australia and New Zealand). Woodland and woodland edge; open moorland in Scotland. Size and weight vary with subspecies and locality, eg: SH 48–50in in *C.e. hanglu*, cwt 167lb in *C.e. corsicanus*, cwt 600–660lb in *C.e. hippelaphus*; females are smaller than males. Antlers branched. varying with subspecies and locality, eg: AL 48in; Pt 20 in *C.e. hippelaphus*; AL 35in; Pt 12 in *C.e. scoticus*. Coat: red, liver or gray brown in summer, becoming darker and

grayer in winter; rump patch yellowish often with a dark caudal stripe; young spotted. Gestation: 230–240 days. Subspecies: 12.

Wapiti
Cervus canadensis
Wapiti or elk (elk in America only).

WN America, Tien Shan Mts to Manchuria and Mongolia, Kansu, China (introduced to New Zealand). Grassland, forest edge; in mountainous regions there is an upward summer movement. SH 51–60in; wt 530–1,000lb females smaller than males. Antlers branched; AL 39in; Pt 12. Coat: in summer, light bay with darker head and legs; in winter, darker with gray underparts; rump patch light colored; young spotted. Gestation: 249–262 days. Subspecies: 13.

Eld's deer E *
Cervus eldi
Eld's deer or thamin.

Manipur, Thailand, Vietnam, Hainan Island, Burma, Tenasserim. Low, flat marshy country. SH 45in; wt? Antlers: the long brow tine and the beam form a continuous bow-shaped curve with forks at the end of the beam; AL 39in (does not include brow tine). Coat: in summer, red with pale brown underparts; in winter, dark brown with whitish underparts; white on chin, around eyes and margins of ears; females lighter colored than males. Young spotted. Walks on the undersides of its hardened pasterns as well as its hooves, a modification to marshy ground. Subspecies: 3.

Sika deer E
Cervus nippon
Sika or Japanese deer.

Japan, Vietnam, Formosa, Manchuria, Korea, N and SE China (introduced to New Zealand). Forest. SH 25–43in; wt 106lb; AL 11–32in; Pt 6–8 depending on race; top points tend to be palmate. Coat: in summer, chestnut-yellowish brown with white spots on sides and a white caudal disc edged with black; in winter, gray brown with spots less obvious; young spotted. Gestation: 217 days. Subspecies: 13.

Rusa deer
Cervus timorensis
Rusa or Timor deer

Indonesian Archipelago (introduced to Australia, New Zealand, Fiji, New Guinea). Parkland, grassland, woods and forest. SH 38–43in male;

34–38in female; wt ? Antlers branched; AL 44in; Pt 6. Coat: brown with lighter underparts and under tail; young unspotted. Subspecies: 6.

Sambar
Cervus unicolor

From Philippines through Indonesia, S China, Burma to India and Sri Lanka (introduced to Australia and New Zealand). Woodland; avoids open scrub and heaviest forest. SH 24–60in; wt 500–600lb. Antlers branched with long brow tine at acute angle to beam and forward pointing terminal forks; AL 24–39in; Pt 6. Coat: dark brown, with lighter yellow brown under chin, inside limbs, between buttocks and under tail; females and young lighter colored than males; young unspotted. Gestation: 240 days. Subspecies: 16.

Père David's deer
Elaphurus davidiensis

Formerly China, never known outside parks and zoos. SH 47in; cwt 297lb. Antlers branched but with tines pointing backwards. AL 31in. Coat: bright red with dark dorsal stripe in summer, iron gray in winter. Tail long. Hooves very wide. Young spotted. Gestation: 250 days.

Subfamily Odocoilinae

Mule deer [E]
Odocoileus hemionus
Mule or Black-tailed deer.

WN America, C America. Grassland, woodland. SH 40in; wt 264lb. Antlers branch dichotomously, ie in arrangement of even forks not as tines from a main beam; Pt 8–10. Coat: rusty red in summer, brownish gray in winter; face and throat whitish with black bar round chin and black patch on forehead; belly, inside of legs and rump patch white; tail white with black at tip or, in Black-tailed subspecies, with black extending up outer surface. Young spotted. Gestation: 203 days. Subspecies: 11.

White-tailed deer [E] [*]
Odocoileus virginianus

N and C America, northern parts of S America extending to Peru and Brazil (introduced to New Zealand and Scandinavia). In North America conifer swamps in winter, woodland edge spring-fall. In South America valleys near water in dry season, higher altitudes in rainy season.

SH 32–40in; wt 40–300lb. Antlers branch from main beam, ie not dichotomously; beams curve forwards and inwards; basal snag longer than in Black-tailed deer; Pt 7–8, fewer in southern races. Coat: reddish brown in summer, gray brown in winter; throat and inside ears with whitish patches; belly, inner thighs and underside of tail white; tail raised when fleeing showing white underside; young spotted. Metatarsal gland on hock shorter (1in) than on Black-tailed deer (3–5in). Gestation: 204 days. Subspecies: 38.

Roe deer
Capreolus capreolus

Europe, Asia Minor, Siberia, N Asia, Manchuria, China, Korea. Forest, forest edge, woodland, moorland. SH 25–35in; wt 37–51lb. Antlers branched; Pt 6. Coat: in summer, foxy red, with gray face, white chin and a black band from the angle of the mouth to the nostrils; in winter, grayish fawn with white rump patch and white patches on throat and gullet; no visible tail, but in winter females grow prominent anal tufts which are sometimes mistaken for tails; young spotted. After the rut in July–August implantation is delayed until December and kids are born in April–June. Gestation: 294 days (including 150 days pre-implantation). Subspecies: 3.

Moose
Alces alces
Moose or elk (elk in Europe only).

N Europe, E Siberia, Mongolia, Manchuria, Alaska, Canada, Wyoming, NE USA (introduced to New Zealand). SH 66–90in; wt 880–1,765lb (females are about 25 percent smaller than males). Antlers large, branched and palmate; Pt 18–20. Shoulders humped, muzzle pendulous, and from the throat hangs a growth of skin and hair—the "bell"—commonly up to 20in long. Coat: blackish brown with lighter brown underparts, darker in summer than in winter; naked patch on the muzzle is extremely small; young unspotted. Gestation: 240–250 days. Subspecies: 6.

Reindeer
Rangifer tarandus
Reindeer or caribou.

Scandinavia, Spitzbergen, European Russia from Karelia to Sakhalin Island, Alaska, Canada, Greenland and adjacent islands (introduced to South Georgia in the South Atlantic).

Many of the reindeer in Europe and Scandinavia are domestic. Woodland or forest edge, all year for some races, but other races migrate to arctic tundra for summer. SH 42–50in male, 37–45in female; wt 200–600lb male. Antlers present in both sexes, smaller and with fewer points in the female; branched with a tendency for palmate top points; the brow points are palmate in the males; males shed antlers November–April, females May-June; AL up to 58in in males; Pt up to 44 in males. Coat: brown in summer, gray in winter, with white on rump, tail and above each hoof; neck paler and chest and legs darker; males have white manes in the rut; only deer with no naked patch on muzzle, which is fur-covered; young unspotted. Gestation: 210–240 days.

Marsh deer [V] [*]
Blastocerus dichotomous

C Brazil to N Argentina. Marshes, floodplains, savannas. SH ?; wt ?; similar in size to small Red deer. AL 24in; the brow tine forks and there is a terminal fork on the beam which gives Pt 8. Coat: rufous in summer, duller brown in winter; lower legs dark; black band on the muzzle; white, woolly hair inside the ears; young unspotted.

Pampas deer [E]
Ozotoceros bezoarticus

Brazil, Argentina, Paraguay, Bolivia. Open, grassy plains. SH 27in; wt ? AL ?; Pt 6. Coat: yellowish brown with white on the underparts and inside the ear; upper surface of the tail dark; hair forms whorls on the base of neck and in the center of back; young spotted. Subspecies: 3.

Chilean huemul [E] [*]
Hippocamelus bisulcus
Chilean huemul or Chilean guemal.

Chile, Argentina, High Andes. SH 36in; wt ? Antlers branched; AL 11in; Pt up to 4. Coat: dark brown in summer, paler in winter; lower jaw, inside ears and lower part of tail white; young unspotted. Ears large and mule-like.

Peruvian huemul [V] [*]
Hippocamelus antisensis
Peruvian huemul or Peruvian guemal.

Peru, Ecuador, Bolivia, N Argentina, High Andes. wt slightly smaller than Chilean huemul. Antlers similar to Chilean huemul but bifurcation closer to coronet. Coat: similar to but paler

than Chilean huemul. Young unspotted.

Red brocket [*]
Mazama americana

C and S America from Mexico to Argentina. Dense mountain thickets. SH 28in; wt 44lb. Antlers simple spikes; AL 4–5in. Coat: red brown with grayish neck and white underparts; tail brown above, white below. Young spotted. Gestation: 220 days. Subspecies: 14.

Brown brocket
Mazama gouazoubira

C and S America from Mexico to Argentina. Mountain thickets, but more open than Red brocket. SH 14–24in; wt 37lb. Antlers simple spikes; AL 4–5in. Coat: similar to Red brocket but duller brown. Young spotted. Gestation: 206 days. Subspecies: 10.

Little red brocket
Mazama rufina

N Venezuela, Ecuador, SE Brazil. Forest thickets. SH 14in; wt ? Antlers simple spikes; AL 3in. Coat: dark chestnut with darker head and legs. Pre-orbital glands extremely large. Young spotted. Subspecies: 2.

Dwarf brocket
Mazama chunyi

N Bolivia and Peru. Andes. SH 14in; wt ? Antlers simple spikes; AL ? Coat: cinnamon-rufous brown with buff throat, chest and inner legs, and white underside to tail; whorl on nape, supra-orbital streak and circumorbital band lacking; smaller and darker than Brown brocket. Tail shorter than in other brockets. Young spotted.

Southern pudu [E] [*]
Pudu pudu

Lower Andes of Chile and Argentina. Deep forest. SH 14–15in; wt 13–18lb. Antlers simple spikes; AL 3–4in. Coat: rufous or dark brown with paler sides, legs and feet, young spotted. Gestation: 210 days.

Northern pudu [I] [*]
Pudu mephistophiles

Lower Andes of Ecuador, Peru, Colombia. Deep forest. Slightly larger than Southern pudu. Antlers simple spikes; AL ? Coat: reddish brown with almost black head and feet. Young spotted. Subspecies: 2.

RAC

Harvest or Harm

Relationship between Fallow deer and man

Fallow deer are among the most familiar of the European deer, perhaps because their elegance has led to their establishment in park herds. Their special appeal to man has dramatically influenced their distribution: they were widespread throughout Europe some 100,000 years ago, but they probably became extinct in the last glaciation, except for a few small refuges in southern Europe. From these relict populations, the spread of Fallow deer to the rest of Europe, their introduction to the British Isles, North and South America, Africa, Australasia and other parts of Eurasia must all have been assisted by man.

In the wild state, Fallow deer are characteristic of mature woodlands; although they will colonize plantations (provided these contain some open areas), they prefer deciduous forests with established understory. These forests need not be particularly large, since they are used mainly for shelter. Fallow deer are primarily grazers, feeding in larger woodlands, on grassy rides or on ground vegetation between the trees, but often leaving the trees to graze on agricultural or other open land.

During the summer, the deer graze out in the open at dawn and dusk. During fall and winter, when grazing is less nutritious, they spend more time within woodland, feeding increasingly on woody browse and coming together in dense concentrations to exploit local areas of abundant mast (beechnuts). Fall is also the breeding season or rut, when bucks establish display grounds in traditional areas within the woodland and call to attract females.

Fallow deer are commonly regarded as a herding species, since they may be seen in concentrations of 40 or more, but in practice social organization is rather complex. The sexes remain separate for much of the year; adult males may travel together in bachelor groups, whereas females and young form separate herds, often in different areas to males. If a woodland is small, it may support only males or only females for much of the year.

Males enter females' areas to establish rutting stands early in fall and then adult groups of mixed sex may be observed through to early winter. Strictly speaking, these large groups are not herds but are usually chance aggregations where numbers of individuals or small groups temporarily coincide at favored feeding grounds. It now appears that the basic social unit is in fact the individual or the mother and fawn. Since the deer occupy overlapping home ranges, temporary associations of 7–14 animals may occur at certain times of year, but these are not of long duration and their composition in terms of individual animals is not constant.

The smaller social groups usually maintain their identity within the large aggregations at feeding grounds. Home ranges are usually comparatively small—50–123 acres (20–50ha) in females and 100–172 acres (40–70ha) in males within the New Forest in Hampshire, England. Home ranges are probably larger in coniferous than in deciduous woodland, and when the deer form associations these too are larger in coniferous areas. The persistence of "herds" of mixed sex after the rut also appears to change with habitat-type.

With their preference for feeding on ground vegetation, Fallow deer in sufficient density can have a significant effect upon their environment. In two experimental areas of mature beech woodland in Hampshire, one occupied by Fallow deer at a density of approximately one per hectare (one per 2.5 acres), the other maintained free of all large herbivores, the grazed area is characterized by an almost total lack of herb-layer, absence of understory shrubs and complete lack of any regenerating tree

► **A tree trunk shredded** by the antlers of Fallow deer in a wood in Worcestershire, England. The damage is caused by the bucks thrashing their antlers in the trees in aggressive display or in clearing the remnants of the protective skin (velvet) from newly grown antlers.

◄ **Fallow deer in bracken.** BELOW during the summer the deer move out of the woodlands at dawn and dusk to graze, returning to the shelter of the woods to ruminate during the middle of the night and in the daylight hours.

▼ **Fallow deer in a glade.** Grassy rides and glades in woodland are especially favored by Fallow deer. Clearings provide good grazing and the surrounding trees protection.

seedlings. This effect is extreme, since the density of deer is unusually high; nonetheless, the impact of the deer can be observed in other ways and in other environments. Where Fallow deer are established in small woodlands in agricultural areas, their habit of moving out into the surrounding farmland to feed may cause significant damage. Within woodland, male deer may inflict considerable damage on individual trees by thrashing them with their antlers, in aggressive display or in cleaning the remnants of the protective skin (velvet) from newly-grown antlers. In certain areas, therefore, where populations are well established, Fallow deer may be a pest both of agricultural and forest crops. As a result, most wild populations are subject to some degree of management for control. Man thus affects not only worldwide geographical distribution of this species, but local abundance too.

More recently, managers have recognized in their deer populations a valuable renewable resource. Instead of management being directed purely towards control, more populations are being harvested for profit. Similarly, the possibility of enclosing populations of Fallow deer and farming them under more intensive conditions as a new agricultural species is being explored. Fallow deer-farms have been established in Germany, Australia, New Zealand and the United Kingdom.   RJP

Moose Foraging

How a moose chooses food

During the summer at Isle Royale, a wilderness in Lake Superior, on the USA–Canada border, moose move daily in a regular manner between aquatic and terrestrial (open-forest) feeding sites; deeper forest areas are used for resting. Their regularity in feeding is reminiscent of the regularity of a person's grocery shopping or a manager's operation of a factory. Both Adam Smith, the founder of the theory of capitalism, and Charles Darwin, the founder of the theory of evolution, noticed this similarity between animal feeding and human economic endeavors. These human endeavors follow a process of decision-making, mathematical or intuitive, ensuring that the groceries provide the best nutrition and/or pleasure and that the factory provides the best profit. Whether moose foraging is truly like such human decision-making—do moose choose the best diet?—was studied at Isle Royale, a 212sq mi (550sq km) site best known for its wolves but also the home of a large moose population.

In the operation of a factory, a manager's actions are limited by available capital, raw materials and labor force, provided the technology is available to convert these quantities into products. A moose in its feeding, likewise, is constrained by daily feeding time, digestive capacity, sodium requirements and energy needs.

A moose's daily feeding time is limited by loss and gain of body heat with the environment; on an average summer day moose can feed on land at air temperatures of 50–59°F (10–15°C) and in ponds from 59–70°F (15–21°C); otherwise it is either too hot or too cold to feed. This time restraint is inflexible but the moose makes the best use of the time available by its ability to crop different plant foods. Cropping rates are set by each food's abundance and by the moose's ability to collect each.

A moose is a ruminant, so its ability to digest food is limited by the amount of time the food remains in the rumen, one of four stomach chambers. The rumen's capacity (volume times emptying rate) imposes another constraint on food intake, and the rumen is filled to differing degrees by each plant food, as each differs in bulkiness.

Moose, like all mammals, require sodium for maintenance, growth and reproduction; at Isle Royale past glaciation has removed soil sodium and oceanic salt spray does not add any. Sodium is certainly in short supply at Isle Royale in the sense that in order to satisfy its sodium needs a moose would have to eat more terrestrial vegetation each day than the rumen could hold. However, aqua-

tic vegetation, such as pondweed, Water lily, horsetails and bladderworts, growing in the beaver ponds, contains a higher sodium concentration. Moose eat aquatics to obtain the sodium, but eating these plants poses problems because they are available to the moose for only 120 days a year when conditions are warm enough for them to feed in the beaver ponds, and because aquatic plants are so bulky that they fill a moose's rumen five times faster than other vegetation. A moose must exceed its personal sodium needs to reproduce. Moose energy requirements for maintenance and growth must also be exceeded to allow reproduction; this is achieved by the ingestion of different plant foods, such as leaves of ash, maple and birch, each with a different energy content.

Given resource limitations, a factory manager seeks to produce a mix of products which maximizes profits. Animals, given their limitations, do not have such a clear goal, but have two possibilities. One goal is to maximize the intake of a nutritional component, generally energy. Maximum energy intake, given feeding limitations, provides a moose with the greatest energy for reproduction in summer and for winter survival, when food is in short supply. The other potential goal is to spend the least possible time in acquiring food. This enables

▲ **A moose browsing on land.** Moose take both leaves from trees and shrubs and also plants from the forest floor. Each day a moose consumes over 8.8lb (4kg) dry weight of terrestrial vegetation or the equivalent of over 20,000 leaves.

▶ **Aquatic foraging in moose** at Yellowstone National Park, USA. In a summer day, a moose will consume 1.9lb (870g) dry weight of sodium-rich aquatic plants, or over 1,100 plants.

▶ **Consumer choice.** Moose must meet their daily energy and sodium requirements, given a limited number of hours in which to feed and a limiting rate at which food can be digested. If the moose's goal is to achieve the maximum intake of energy, it will spend just long enough feeding on aquatics to satisfy its sodium requirement, and spend the rest of the time feeding on terrestrial vegetation, up to its digestive capacity. If the moose wishes to spend the minimum time in feeding (so that it can devote time to reproductive behavior), it will take the minimum of terrestrial vegetation and fill up with easily obtained, bulky aquatic vegetation. Observations of moose foraging show that they take the first course.

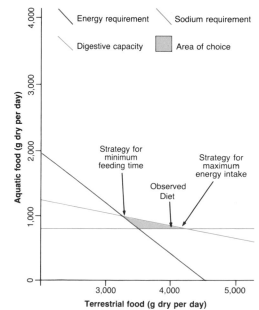

a moose to have more time for non-feeding activities, such as caring for young, and reduces exposure to hostile environmental factors, such as predators or excessive heat loss etc.

Over three years, measurements of moose physiology and behavior, and the summer environment, were made in a 1.2 sq mi (3 sq km) area at Isle Royale. A mathematical model of moose feeding was constructed with these data to predict what a moose should eat to satisfy either of the two potential goals. The model is identical to models used in industry to predict output!

Comparing a moose's observed diet with the model's predictions indicates that moose maximize energy intake. The diet contains enough aquatic vegetation to supply sodium demands, including sodium storage for times when aquatic plants are unavailable,

while terrestrial plants are sought for energy. Therefore, moose appear to make foraging decisions in the same way that decision-making occurs in human economic endeavors.

Why should a moose maximize energy intake? Evolution should have produced a moose diet which copes best, given limitations, with environmental factors controlling reproduction and survival. Because of mooses' large size, predation by wolves probably does not limit their survival and reproduction, with only young, old or unhealthy animals vulnerable to wolves. If such predation were important, moose might have adopted a stategy that minimized feeding time. Rather, the observed energy-maximized diet indicates that food limits moose survival and reproduction.

GIRAFFE AND OKAPI

Family: Giraffidae
Two species in 2 genera.
Order: Artiodactyla.
Distribution: Africa.

Giraffe

Giraffa camelopardalis
Distribution: Africa south of the Sahara.
Habitat: open woodland and wooded grassland.
Size: head-body length (male 12–15ft
(3.8–4.7m); tail length (excluding tassel) (male)
31–39in (80–100cm); height to horn tips
(male) 15–17ft (4.7–5.3m), female 13–15ft
(3.9–4.5m), weight (male) 1,765–4,255lb
(800–1,930kg) (female) 1,213–2,601lb
(550–1,180kg).
Coat: patches of variable size and color (usually
orange-brown, russet or almost black)
separated by a network of cream-buff lines.
Gestation: 453–464 days.
Longevity: 25 years (to 28 in captivity).
Subspecies: **West African giraffe** (*G.c. peralta*),
Kordofan giraffe (*G.c. antiquorum*), **Nubian
giraffe** (*G.c. camelopardalis*), **Reticulated giraffe**
(*G.c. reticulata*), **Rothschild giraffe** (*G.c.
rothschildi*), **Masai giraffe** (*G.c tippelskirchi*),
Thornicroft giraffe (*G.c. thornicrofti*), **Angolan
giraffe** (*G.c. angolensis*), **South African giraffe**
(*G.c. goraffa*).

Okapi

Okapia johnstoni
Distribution: N and NE Zaire.
Habitat: dense rain forest; prefers dense
undergrowth in secondary forest. Often
recorded in forest beside streams and rivers.
Size: head-body length 6–7ft (1.9–2m); tail
length (excluding tassel) 12–17in (30–42cm);
height 5–6ft (1.5–7m); weight 465–550lb
(210–250kg).
Coat: velvety dark chestnut, almost purplish
black, with conspicuous transverse black and
white stripes on hindquarters, lower rump, and
upper legs; the female is redder in color.
Gestation: 427–457 days (in captivity).
Longevity: 15 or more years in captivity).

THE giraffe is often referred to as "the
animal built by a committee," an assem-
blage of left-over parts, put together after the
divine creator had run dry of ideas. This
description belittles an animal that is
superbly adapted for feeding on the high
foliage beyond the reach of other hoofed
mammals, and which is one of the more
widespread and successful herbivores of the
African savanna.

The curious anatomy of the giraffe is
accentuated by the short length of the body
in relation to the pronounced length of the
neck. This foreshortening is exaggerated
by the height of the legs, the forelegs being
longer than the hind, so that in profile the
animal slopes continuously from its horn
tips to its tail. The neck has the usual seven
vertebrae of most mammals, although each
is greatly elongated. The thoracic vertebrae,
especially numbers four and five, have large
forward-facing dorsal spines which form the
conspicuous shoulder hump, and serve as
anchors for the attachment of the large
muscles that support the head and neck.
The tail has a 3-foot long terminal tuft of
black hairs, producing an effective whisk
against the tsetse fly. The soup-plate-sized
hooves are used as offensive weapons,
usually in the defense of the calves, for the
powerful kick from the front feet can kill a
lion. The hide is thick for defense.

A unique feature of giraffes is the progressive laying down of bone material
around the skull, particularly of bulls, which
in late life may be so covered with bony
lumps and concretions that the original

profile is obliterated. These bony growths
are typically sited above each eye socket,
centrally in the forehead region, and occasionally at the back of the skull, producing
unusual "three-horned" and "five-horned"
giraffes. An adult male skull 33lb (15kg)
may weigh three times as much as a
female skull 10lb (4.5kg) which has few
bony growths.

The giraffe is one of the few ruminants
born with horns. At birth, the cartilaginous
cores lie flat upon the skull, but adopt the
typical upright stance during the first week
of life. Both sexes have horns which are
unusual in being covered with skin, with a
terminal tuft of black hairs. The horns start
to become bony as soon as they are upright.
The horns of the males are thicker and
heavier, often fusing at their base and losing
the terminal hair tuft. The horns and the
head are used in fights between males.

Other peculiar anatomical features include the canine teeth, which are splayed out
into two or three lobes to comb the leaves off
shoots, and the long black tongue, which
can be extended up to 18in (46cm) and is
used to gather food into the mouth. To
compensate for the sudden increase in blood
pressure when the head is lowered, the
giraffe has very elastic blood vessels and
valves in the venous system of the neck.

Coat pattern and color are very variable,
ranging from the large regular chestnut
patches separated by a network of narrow
white lines in the Reticulated giraffe, to
the very irregular, star-shaped or jagged
patches of the Masai gifaffe, which may vary

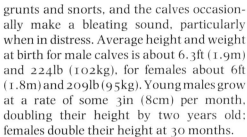

▷ **On the run.** OVERLEAF Giraffes do not form stable herds but they do band together in loose groups for protection against predators. Giraffes are extremely mobile and their long-range visual acuity is excellent, enabling groups several miles apart to be in communication.

◀ ▼ **Browsing styles.** The sex of a distant browsing giraffe can be reliably determined by stance, the female LEFT bending over the vegetation and the male feeding at full stretch BELOW. This reduces competition between the sexes.

in colour from pale orange to almost black. Horn growth and coat pattern are dubious characteristics by which to classify giraffes, although they have been frequently used as such. Each giraffe has its own unique pattern of coat markings, like human fingerprints, which enables it to be distinguished from all other giraffes. The pattern remains constant from birth to death, although the color may change, so the animal can be recognized throughout its life.

The best developed sense of the giraffe is its excellent sight, although both hearing and the sense of smell are acute. Though usually silent, they can produce a variety of grunts and snorts, and the calves occasionally make a bleating sound, particularly when in distress. Average height and weight at birth for male calves is about 6.3ft (1.9m) and 224lb (102kg), for females about 6ft (1.8m) and 209lb (95kg). Young males grow at a rate of some 3in (8cm) per month, doubling their height by two years old; females double their height at 30 months.

The current distribution of giraffes is confined to the open woodlands and wooded grasslands of Africa south of the Sahara. They are typically associated with acacia (*Acacia*), myrrh (*Commiphora*), *Combretum* and open myrobalan (*Terminalia*) woodlands. Their distribution in the miombo (*Brachystegia*) woodlands of central and southern Africa is largely confined to the drainage lines where *Acacia* and *Commiphora* are common. They are absent from the central African rain forest.

Giraffes are exclusively browsers, although in the absence of browse in captivity, they will graze, often when lying down. They are highly selective feeders, the bulk of their diet comprising the leaves and shoots of trees and shrubs, supplemented by climbers, vines and some herbs. Flowers, seed pods and fruits are also eaten in season. Such material is rich in nutrients and does not show the same marked decline in quality as do the grasses during the drier months of the year. By shifting their feeding in the dry season into the woodlands beside streams and rivers, which continue to produce new leaf and shoot material throughout the year, giraffes are able to exploit a high-quality diet, and thus to breed, at all times of the year. They can also escape the seasonal limitation on breeding imposed upon most grazers by nutritional stess.

How the food is gathered depends upon the plant's defenses. With the acacias, protected by an array of long spiky and short recurved thorns, individual leaf whorls and shoot tips are bitten off, the tongue and sensitive lips being used to gather the material into the mouth. The roof of the mouth is heavily grooved, and this, together with the copious production of viscous saliva, enables the giraffe to compress and swallow the thorny food. With the thornless broad-leaf shrubs, individual shoots are pulled through the mouth, the leaves being combed off by the lobed canine teeth.

Giraffe cows feed for some 55 percent of each 24 hours, bulls for about 43 percent. Daytime feeding occurs mainly in the three-hour periods after dawn and dusk, with an increase in ruminating, or chewing the cud, during the hot midday period. Nocturnal

behavior depends upon the amount of moonlight: more activity, including feeding, occurs during bright nights. Adult bulls each day consume some 42lb (19kg) dry weight, 145lb (66kg) fresh weight, cows some 35lb (16kg) dry, 128lb (58kg) fresh. Expressed as a proportion of body-weight, cows (2.1 percent per day) have a greater relative rate of food intake than bulls (1.6 percent). Although they can survive without water for relatively long periods, giraffes make regular visits to water, often trekking over long distances. Giraffes can rest standing, but they often lie down with their legs folded beneath them. The neck is held vertical except during short periods of sleep, usually of about five minutes' duration, when the head is rested on the rump.

A characteristic feature of giraffe feeding is the vertical separation between bulls and cows. The typical feeding stance of an adult male is with the head and neck at full vertical stretch, often with the tongue extended, to reach the shoots on the underside of a mature tree canopy, up to 19ft (5.8m) above ground level. The typical stance of an adult female is with the neck curled over, feeding at body- or knee-height upon regenerating trees and low shrubs. So characteristic is this sexual disparity that a giraffe can be reliably sexed from its feeding behavior alone at distances too great for other characteristics to be seen.

Locomotion is peculiar in that the "pacing" walk of the giraffe involves swinging the two legs on the same side of the body forward at almost the same time. The left foreleg leaves the ground soon after the left hindleg has begun its swing forward. When galloping, the hindlegs are brought forward almost together and placed outside the front. Maximum galloping speed is 31–37mph (50–60km/h). Calves, having less inertia, sprint ahead of the adults.

Female giraffes conceive for the first time in their fifth year (48–60 months old), although sexual maturity may be delayed for a year or more by poor nutritional conditions. With a gestation period of 15 months, a mean interval between births of some 20 months, and a maximum longevity of 25 years, a cow may produce up to 12 calves in her lifetime, although about half that number is probably the average. Twin fetuses have been recorded at autopsy, but the live birth of twins is extremely rare. Males become sexually mature at about 42 months, but do not have the opportunity for successful mating until fully grown, at 8 or more years old. Sexual libido declines in very old males. Although early in courtship, subordinate bulls may consort with a receptive female, a succession of bulls of progressively increasing dominance status will temporarily consort with the cow. By the time she will stand to be mated, some 24 hours

Traditional Calving Grounds

Giraffes do not drop their calves just anywhere in the bush, for births are concentrated in discreet calving grounds. The home range of an adult female may include several such calving grounds, yet their total area probably accounts for less than 10 percent of the animal's range. From a sample of 89 calves born in the Serengeti Natonal Park, only five were located outside the calving areas.

Calving grounds show no constant feature, some comprising open grassy hilltops, others dense riverside thickets of young trees. Anecdotal evidence suggests that an area continues to be used for births for a relatively long time (at least 15 years), despite major habitat changes. A calving ground in the Serengeti has changed from open mature acacia parkland to dense regeneration thicket as a result of elephant activity, yet it is still used by giraffes as a birth area.

A probable explanation is that the concentration of births facilitates the formation of calf groups or creches, especially for calves born outside the main birth peaks. A detailed familiarity with the calving area may also have some survival value against predators at this time of extreme vulnerability.

A female giraffe is faithful to a specific calving ground. From a sample of 25 cows in the Serengeti, who were recorded as giving birth to second offspring, all the second births occurred in the same calving ground as the first. One particular female, with a home range of some 37sq mi (95sq km), dropped her second calf within 50ft (15m) of the birth-site of her previous calf born 18 months earlier. Surprisingly, the predators do not appear to have appreciated the significance of these calving grounds: if they did, dispersed births would become a more effective antipredator strategy.

Evidence from long-term radio-tracking suggest that female home ranges may not be static but drift. This drift may result in the calving ground patronized by a particular female moving out of her central core area into her buffer zone, and even out of her home range altogether. This explains the curious behavior of some heavily pregnant cows, particularly old females, who suddenly desert their usual haunts to give birth in a distant calving ground, despite the presence of a choice of well-used calving areas within their normal home range.

into heat, the most dominant bull of the locality is courting the female. A relatively small number of top bulls do all the mating.

Calves are born in isolation. The use by the cows of traditional calving grounds facilitates the formation of calf groups, which the newborn joins at about 1–2 weeks old. It is not unusual to see groups of very young calves, some still with the umbilical stump, apparently abandoned by their mothers in the middle of the day. In fact, the collective vigilance of these groups is very acute, and the predators are largely inactive in the heat of the day. The cows benefit by visiting distant feeding areas without having to spend time on the care of their offspring, resulting in good lactation.

Female giraffes are excellent mothers, and will vigorously defend their offspring against predators, kicking out with their feet. Lions are the principal predator, although newborns up to about three months old may also be killed by hyenas, leopards and even African wild dogs. In the Serengeti, the first-year calf mortality is about 58 percent. Some 50 percent of calves die in their first six months of life, with mortality greatest during the first month (22 percent). Mortality in the second and third years drops to about 8 percent and about 3 percent per annum in adults. Yet despite this apparently high rate of calf mortality, the giraffe population of the Serengeti is increasing at some 5–6 percent per annum. Male calves are weaned at about 15 months, female calves a couple of months later. No differences in the mortality of male and female calves have been observed.

Giraffes form scattered herds, the compositions of which are continually changing. In over 800 consecutive daily observations of an adult female in the Serengeti, the herd composition remained unchanged over the 24-hour period on only two occasions. The individual is thus the social unit in giraffe society, but the giraffe being a gregarious animal, individuals band together into loose groups for protection against predators. These loose associations are an adaptation to the mobility and extreme long-range visual acuity of giraffes, for groups several miles apart may be in visual communication.

Home ranges in giraffes are large, about 46sq mi (120sq km) for adult cows, but smaller in mature bulls, and larger in young males, who are great wanderers. Not all the range is equally used, for about 75 percent of sightings are confined to the central 30 percent of the area. This central core probably supplies all the requirements for day-to-day survival, but is surrounded by a familiar area which buffers the individual from the unknown world. Behavioral differences within the two areas are conspicuous, with more feeding in the central core and more alert vigilance and greater mobility in the buffer zone.

In areas where their home ranges overlap, individuals may associate, forming loose groupings. However, associations between certain individuals occur with a much higher frequency than between others, implying that giraffes have friends. This in turn suggests that they can recognize each other. Observations of their social behavior, particularly the investigation of unknown animals in their buffer zones, supports this suggestion. The exceptionally high frequency of association between pairs of adult cows may represent mother-daughter relationships.

▶ **Mating giraffes.** Males spend much of their time patrolling, looking for females in heat. A dominant male will mate with all such females he encounters.

▶ **A giraffe calf** BELOW at 5–7 days old. After weaning, young females remain within the home ranges of their mothers. Young males band together into all-male groups, and disperse from the home range where they were born in their third or fourth year.

▼ **Necking giraffes.** "Necking" is ritualized fighting indulged in mainly by young bulls to determine dominance. The necks are slowly intertwined, pushing from one side to the other, rather like a bout of arm-wrestling in humans.

Bulls are non-territorial, and amicably coexist together within overlapping home ranges. The reason for this harmony is the dominance hierarchy, the predominant feature of the bull's society. Each individual knows its relative status in the hierarchy, which minimizes aggression. Status is reinforced at encounters between two bulls by threat postures—standing tall to impress the rival. Cows also have a dominance hierarchy, although its manifestation is more subtle, usually being confined to the displacement of subordinate females at attractive feeding sites.

Young bulls have developed an elaborate ritual, called necking, to determine dominance. Necking is ritualized fight in which the two participants intertwine their necks in a slow-motion ballet. Its purpose is to assess the relative status of the two individuals. Occasionally, the intensity of the encounter may increase and the two combatants, standing shoulder to shoulder, exchange gentle blows upon each other's flanks with the horn tips. Such bouts are interspersed with much pushing, but usually degenerate into necking, which typically terminates with one animal enforcing his dominance by mounting the other.

Serious fighting, when sledgehammer blows are exchanged using the side of the head, is rare, being confined to occasions when the dominance hierarchy breaks down. This usually occurs with the intrusion of an unknown nomadic male that has no position in the local hierarchy. To protect the brain when fighting, the entire roof of the skull is pocketed with extensive cranial sinuses.

Juvenile females, after weaning, remain within the home ranges of their mothers. Juvenile males band together into all-male groups, and disperse from their natal areas in their third or fourth year. This is the age of most necking encounters, when the hierarchy is established. Top bulls can exert their dominance only over the relatively small core area of their home ranges, for their status declines as they move out into their buffer zones. There thus exists some latent territoriality, in that the dominance status has some geographical connotation. Maybe ancestral giraffes were territorial.

Within its core area, a dominant bull will mate with most of the cows that come into heat, provided that he can find them. Bulls therefore spend much of their time patrolling their core areas, reinforcing their dominance over other bulls, and seeking out females coming into heat. The bull's life-strategy is to feed for the minimum time necessary to attain the nutritional requirements for reproduction, leaving the maximum time for dominance assertion and finding cows. On the other hand, the female strategy is to feed for as long as possible so as to maximize her nutritional intake, to ensure year-round breeding and maximum reproductive success.

The constant searching for females results in the top bulls leading somewhat solitary lives. On sighting a group of giraffes, a mature bull will join the heard, testing each female in turn by sampling her urine with the flehmen, or lip-curl, response. If none of the females is in heat, the bull will soon leave the group to continue his solitary searching elsewhere. If a female is in reproductive condition, the bull will consort with the cow, displacing the subordinate bull already courting her.

Giraffes are still relatively common in East and South Africa, although their distribution in West Africa has been fragmented by poaching. In their stongholds, such as the Serengeti, they may reach densities of 3.9–5.2 animals per sq mi (1.5–2 per sq km) making an important contribution to the total animal biomass of the savanna. At such densities, their browsing can exert a significant brake upon the development of tree regeneration. In areas where elephants are destroying mature trees, this retarding of the growth of the young replacements can present a serious management problem. The problem is particularly acute in areas of high fire frequency, where heavy browsing prolongs the period that the regeneration is vulnerable to fire. A combination of burning and browsing can jeopardize the conservation of the woodlands, especially in areas where the existing mature canopy is being eliminated by elephants. But because giraffes do not actually kill trees, and merely act as a catalyst exacerbating the impact of other agents of mortality, the most effective solution to the problem is to remove the real tree killers, especially fire.

Giraffe meat has for long been sought by subsistence hunters. Certain tribes, such as the Baggara of southern Kordofan in Sudan, the Missiriés of Chad, and the Boran of southern Ethiopia, have traditionally hunted giraffe from horseback, the quarry receiving a special reverence and being of particular social significance to these tribes. But this mystique has not prevented local exterminations of giraffe by over-hunting. The tourist trade in giraffe-hair bracelets has

also encouraged poaching, only the tail being removed from the carcasses.

Giraffes have important potential as a source of animal protein for human consumption. Because they feed upon a component of the vegetation that is unused by existing domestic animals, except possibly the camel, they do not compete with traditional livestock. Cattle ranchers and pastoralists have therefore been tolerant of giraffes, although they can damage stock fences. Giraffes rapidly become used to the presence of man and, as a free-range resource, can be easily cropped. At high densities, it has been estimated that giraffes could provide up to one-third of the meat requirements of a pastoral family, by exploiting the high browse which is currently neglected. This would reduce the dependence upon the sheep and goats that are responsible for much of the overgrazing and erosion problems in Africa today.   RAP

▶ **The elusive okapi,** which looks somewhat like a cross between a giraffe and a zebra. Okapis are secretive, solitary animals, relying upon their acute senses of hearing and smell to evade predators; unlike giraffes, their sight is relatively poor. Okapis are generally described as being nocturnal but in undisturbed areas they may be active in the day. They are pure browsers, feeding upon the leaves of young shoots of forest trees. Seeds and fruit have also been recorded in their diet.

▼ **To drink,** giraffes have to splay their forelegs very widely and then bend the knees.

open savanna in the giraffe, forest in the okapi. Early reports suggested that okapis, especially the males, were nomadic. However, their regular use of tracks linking favorite feeding areas suggests a more sedentary way of life. Unlike giraffes, okapis have glands on their feet, and they have also been reported marking bushes with urine. These factors, together with their solitary existence, imply a social system based on male territoriality.

In captivity, heat lasts for up to a month, during which time the male consorts closely with the female. Females advertise their condition by urine marking and by calling, although at other times okapis are silent. The exceptionally long period of heat may be to ensure sufficient time for the male to locate the female and to overcome their solitary inclinations before mating. The initial approach and contact phases of okapi courtship are marked by more female aggression and male dominance display than in giraffes, presumably to overcome the defensive reactions of typically solitary individuals. Male courtship behavior is more antelope-like, including flehmen (lip-curl), display of the white throat patch, leg kicking and head tossing. Male encounters, particularly in the presence of a female in heat, are often aggressive, the ritualized neck fight being reinforced by periodic charging and butting with the horns.

Calves are born with non-adult proportions, having a small head, a short neck and thick long legs. The conspicuous mane is much reduced in the adults. Calves remain hidden for the first few weeks. Vocalization is important to maintain contact between mother and offspring. In captivity, calves are weaned by six months. The horns of males develop between the first and third years. Leopards are the principal predator of calves. Like giraffes, females defend their offspring by kicking with their feet. The strong contrast of the zebra-like stripes of the back legs and flanks are probably important for calf imprinting and as a follow-me signal.

The okapi has been protected by government decree since 1933. Although the Zaire authorities have made real efforts to enforce this protection, it is difficult to prevent hunting in remote forest areas, particularly when the meat is so prized. Subsistence hunting by pygmies is unlikely to endanger the status of the species. This is not the case, however, with intensive commercial poaching, which now threatens all the larger forest mammals, including the okapi, in the more accessible areas. RAP

The **okapi** remains one of the major zoological mysteries of Africa, so little is known about its behavior in the wild. The species was only officially "discovered" in 1901 by Sir Harry Johnston, the British explorer, whose interest had been aroused by the persistent rumors of a horse-like animal, living in the forests of the Belgian Congo, that was hunted by the pygmies. Okapi is the name that the pygmies have given to this creature.

Its present distribution is confined to the rain forest of northern Zaire, between the Oubangi and the Uele rivers in the west and north, and up to the Uganda border and the Semliki river in the East. Like most forest-dwelling large mammals, it is very localized, but in areas of suitable habitat is relatively common, particularly in the Ituri and Buta regions. Local population densities of 2.6–6.4 per sq mi have been suggested. Suitable habitat comprises dense secondary forest where the young trees and shrubs increase food availability. Riverside woodland, and particularly clearings and glades where the penetration of light encourages the production of low browse, is favored. Closed high-canopy forest with little ground vegetation is not suitable habitat.

Characteristic giraffe-like features include skin-covered horns, lobed canine teeth, and an extendable long black tongue which is used to gather food into the mouth. Only males are horned although females sometimes have a variable pair of horn sheaths.

Differences in behavior and ecology between the two giraffid species arise from differences in the habitat in which they live:

PRONGHORN

Antilocapra americana
Pronghorn, prongbuck, antelope, berrendos
(Mexican), Tah-ah-Nah (Taos), Te-na (Piute),
Cha-o (Klamath).
Subfamily: Antilocaprinae.
Family: Bovidae.
Order: Artiodactyla.
Distribution: W USA and Canada, parts of
Mexico.

Habitat: open grass and brushlands, rarely
open coniferous forests; active in daytime to
twilight.

Size: head-to-tail length 55in
(141cm); shoulder height 34in
(87cm); weight 103–154lb
(47–70kg); horn length
(female) 1.5in (4.2cm); (male)
17in (43cm). Females fractionally
smaller than males.

Coat: upper body tan; belly, inner limbs,
rectangular area between shoulders and hip,
shield and crescent on throat, and rump white.
Face mask and subauricular gland on mature
males black.

Gestation: 252 days.

Longevity: 9–10 years (to 12 years in
captivity).

Subspecies: *A.a. americana, A.a. peninsularis* [E].
A.a. mexicana, A.a. sonoriensis [E].

[E] Endangered.

THE pronghorn is renowned in folklore
and in fact for its speed, endurance and
curiosity. Speeds of up to 55mph (86km/h)
have been recorded and 45mph (70km/h)
can be maintained for over 4mi (6.4km).
They will approach and inspect moving
objects, even predators, from long distances.
Early settlers used this "curiosity" to attract
pronghorns within gunshot range by flagging (tying handkerchiefs to poles and waving them in the air). Flagging is now illegal,
but contributed to the pronghorn's decline
from 40–50 million in 1850 to only 13,000
by 1920.

Pronghorns have chunky bodies with
long, slim legs. Their long, pointed hooves
are cloven and cushioned to take the shock
of a stride that may reach 27ft (8m) at a full
run. Both sexes have black horns that are
shed and regrown annually. The buck's
horns have enlarged forward-pointing
prongs below backward-pointing hooks.
The doe's horns are rarely long enough to
develop prongs or hooks. Pronghorn eyes
are unusually large and are set out from the
skull, allowing a 360° field of vision. Long,
black eyelashes act as sun-visors. The
beautifully marked tan and white body and
neck are unique among North American
large mammals. These markings are related
to the pronghorn's use of open prairie and
brushlands and are important in communication. Bucks have a black face mask and
two black patches beneath the ear. These
black markings are used in courtship and
dominance displays. Scent is an important,
but little understood aspect of pronghorn
communication. Males have nine skin
glands (2 beneath the ears, 2 rump glands, 1

above the tail, 4 between the toes) and does
six (4 between the toes, 2 rump). Rump
glands produce alarm odors, and scent from
the glands beneath the ears is used to mark
territories and during courtship.

The pronghorn is the only living species of
the subfamily Antilocaprinae which in the
late Pleistocene (about 1.5 million years
ago) had 5 genera and 11 species. The
subfamily is North American in origin and
distribution. Current subspecies differ
slightly in size, color, and structure, with
darker, larger forms in the north and paler,
smaller forms in the south.

The pronghorn diet varies with the local
flora, but on a plant-type basis they eat:
forbs (herbs other than grass), shrubs,
grasses, and other plants (cacti, domestic
crops etc). Forbs are eaten from spring to
late fall and are critical to good fawn
production. Shrubs are eaten all year, but
are most important in winter. Grasses are
used in spring and other food types vary
locally in importance. Succulence is an
important criterion in deciding food choice,
reproduction, timing of migrations, and
placement and quality of territories. Pronghorn teeth are adapted to selective grazing
and they grow continually in response to
wear and provide an even surface for grinding rough vegetation.

Both sexes mature at 16 months, but does
occasionally conceive at 5 months. Only
dominant males breed and this usually
delays a male's first breeding until 5 years of
age or older. Twins are common on good
range and females produce 4–7 eggs. Up to
four fertilized eggs may implant and when
this occurs the tip of the embryo in the upper
chamber of the two-horned (compartmented) womb grows and pierces the fetal
membranes of the embryo lower in the
womb, eventually killing its sibling while
still in the womb.

The rut occurs from late August till early
October, with exact timing depending on
latitude; it only lasts 2–3 weeks in any given
area. Fawns are born in late May to early
June and on good range weigh from
7.5–8.6lb (3.4–3.9kg) at birth. At two days
old a fawn can run faster than a horse, but it
lacks the stamina needed to keep up with a
herd in flight; therefore it hides in vegetation
until 21–26 days old. Fawns interact with
their mothers for only 20–25 minutes each
day; even after an older fawn joins a nursery
herd this pattern continues. Does nurse,
groom, distract predators from their fawns,
and lead their fawns to food and water.
Nursing and grooming continue for about
4–5 months, but development of aggressive

▲ **Pronghorns in open prairie.** There were perhaps 50 million pronghorns on the plains of North America before the arrival of settlers.

◄ **Lying out.** Newborn pronghorns hide until they are able to move with the herd.

▼ **A male pronghorn,** showing the curiously hooked horns. They are shed after the rut and the males join the female herds.

behavior may cause weaning of males 2–3 weeks earlier than females. Adult bucks show no direct paternal care.

Pronghorns and Roe deer are the only two territorial species of ungulates found in northern latitudes. Males defend and mark territories from early March to the end of rut even though mating only occurs in the fall. Territories may range in size from 0.09–1.68sq mi (0.23–4.34sq km). Doe-fawn herd home ranges usually include several territories and may range in size from 2.45–4.05sq mi (6.35–10.50sq km). Bachelor herd home ranges are located adjacent to territorial areas and are inferior to territorial areas in forage quality. Bachelor home ranges may vary in size from 2–5sq mi (5.1–12.9sq km).

Over 95 percent of all mating is on territories by territory holders. North American rangelands have relatively continuous forage distributions, but within these ranges there are pockets of better forage that a male can defend from rivals. These more productive areas occur in depressions where soils are more moist and richer in nutrients. There is a strong correlation between a male's dominance position established as a bachelor and the quality of the territory he acquires. Does mate most often with males on the best territories and may return to mate on the same territory throughout their lifetime. Therefore, 30–40 percent of all mating may occur on only the best one or two territories in a given area. A top male on a high-quality territory might hold it up for 4–5 years before being replaced, and might sire from 15–30 percent of all fawns for a given year. In a given age-class of bucks, as many as 50 percent may never reproduce.

Early conservation efforts have brought pronghorns back from near extinction to a population of over 450,000. While this is low compared to their former abundance, it is a significant recovery and provides excellent recreational opportunities for hunters, photographers and nature watchers. However, oil exploration and strip mining for coal now pose threats to their habitat.

DWK

WILD CATTLE AND SPIRAL-HORNED ANTELOPES

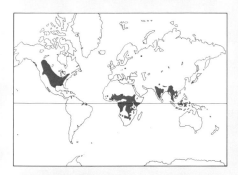

▶ ▲ **The four-horned antelope** RIGHT and the
nilgai ABOVE are the two remaining species of
the primitive tribe Boselaphini from which
modern cattle arose.

WHEN we think of cattle, our minds turn to peaceful herds of dairy cows or to placid beasts of burden. These are reasonable impressions, since there are 1.2 billion Domestic cattle in the world and 130 million water buffalo. In contrast, the kouprey, a very rare forest ox of Cambodia, probably numbers only a few dozen, if indeed it is not already extinct.

Cattle (tribe Bovini) evolved from animals resembling the present-day Four-horned antelope and the nilgai (tribe Boselaphini). The Bovini achieved great diversity in the warm Eurasian plains of the Pliocene (5–3 million years ago) and, in the form of the yak and bisons, they evolved into cold-tolerant species which could cope with the rapid climatic changes of the Pleistocene. Only the bison made the passage via Siberia and Alaska to the New World, spreading as far south as El Salvador.

The challenge of the Pleistocene was also met by the highly successful aurochs, ancestor of our cattle, which spread out from Asia, and after the last Ice Age occupied a vast range from the Atlantic to the Pacific, and from the northern tundras to India and North Africa. A distinct form of aurochs inhabited India, Asia and Africa. As agriculture spread, numbers declined and the last one died in 1627.

The banteng, the gaur, the yak and the Water buffalo have also been domesticated, and with domestication came experimentation. Hybrids between cattle and yak are economically important in Nepal and China. Attempts have been made in North America

to develop a breed (called the cattalo) based on a bison-cattle cross. However, the male progeny of these hybrids are infertile.

In many places, domesticated animals have been allowed to revert to the wild state. In the Northern Territory of Australia, perhaps as many as 250,000 water buffalo roam free, descendants of animals left behind when military settlements were abandoned in the 19th century. The nilgai, although never domesticated, has been successfully introduced to Texas, where it lives as a wild animal. In Britain, many stately homes and castles had parks with herds of more-or-less wild cattle. The most ancient such herd is the Chillingham herd in Northumberland, England.

All species of wild cattle have a keen sense of smell which they use to detect enemies, communicate information to each other and find food—while grazing, they constantly sniff the pasture. Sight and hearing are generally good though not particularly

Domestic cattle; their rumen contents ferment at 104°F (40°C). This "central heating system" means that some breeds do not have to generate extra heat (by shivering, eating more etc) even at 0.4°F (−18°C), in still dry air. The environment of the yak or the bison can be very rigorous, and perhaps this is one reason why the embellishments of the bull bison and bull yak are hairy and heat-conserving (ie the beard and the long fringes on the body) rather than fleshy and heat-radiating (ie the dewlaps and dorsal ridges of the Asian wild cattle).

The Boselaphini have what is thought to be a primitive form of social behavior. The Four-horned antelope lives a solitary life but forms small groups during the rut, probably marking out territories with its facial and foot glands. Nilgai cows and calves tend to stay in herds. The bulls are solitary except during the rut, when they establish territories and gather breeding herds of up to 10 cows, fighting other bulls with their horns. Both species advertise their presence by defecating in particular places.

The Bovini, in contrast, have developed the herd life-style. Little is known of the anoas and the tamarau. In all the other species, females spend their lives in groups of stable composition organized on the basis of personal recognition and which consist of cows, their offspring and perhaps further generations. Young bulls generally leave at about 3 years. These groups are small (ie up to 10) in the gaur, banteng and African buffalo (Forest subspecies). The American and European bison usually live in larger groups (10–20). In all these species, temporary aggregations may form particularly in the breeding season. The African buffalo (Cape subspecies) lives in very large herds. In the open plains and wooded grasslands of the Serengeti, they average 350. Feral Water buffalo in Australia also live in an open habitat; their herds average 100–300, made up of family groups of up to 30 closely related animals. Bulls live alone or in small bachelor herds numbering up to 10–15 which join the female groups during the breeding season or, in the case of the Cape buffalo, when the rains start. In species which inhabit relatively dense cover (Wild water buffalo, banteng, gaur, European bison and forest-dwelling populations of the American bison), individual mature bulls have been seen in cow groups all year.

Males need to be able to assess each other's fighting potential and so avoid dangerous fights between adversaries of comparable status. Their facial musculature does not permit a wide range of expressions,

▲ **Yaks in harsh terrain** near Khumjung, Tibet. Cattle thrive in such extremes. maintaining their fermentation system in the rumen at 104°F (40°C).

▷ **The Viking helmet look** OVERLEAF of the African buffalo, seen here among Red oat grass and attended by oxpeckers.

acute. Domestic cattle have partial color vision—they cannot distinguish red. The sense of taste and of touch on the lips are important in food selection, and the more succulent plant parts are always preferred.

When not feeding, the way that the Bovini spend their time is often determined by the need to avoid heat and insects. Although all species sweat, they frequently have to seek shade. Water buffalo are particularly sensitive to heat, and depend on wallowing; in many places, they are only active at night. Sleep has been studied in Domestic cattle—bouts typically last 2–8 minutes and total no more than one hour in 24.

Cold weather presents few problems to

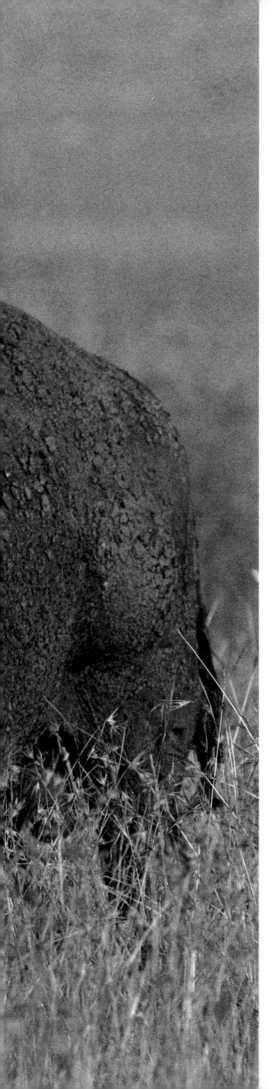

so posture and movement are used instead. The few contests which escalate to actual combat may result in serious injuries. The horns of bovids are very effective anti-predator defenses. A herd of Cape buffalo is very likely to attack a lion; this is such an effective defensive unit that blind, lame or even three-legged individuals may continue to thrive within it. Tigers, however, frequently kill fully grown gaur, and solitary African buffalo bulls often fall victim to lions. Evidently the habit of associating in herds is a vital part of their defense strategy.

The alarm call of the Bovini is an explosive snort, quickly followed by the alarm posture in which the head is held high, facing the danger, with the body tensed. When running away, gaur have the unusual habit of thumping the ground with their forelegs in unison. The Cape buffalo's alarm call brings the whole herd to the defense of the frightened animal.

Animals living in cohesive herds have to coordinate their activities: among Cape buffalo the whole herd may change activity within a few minutes, for example from grazing to lying down, and the American bison's habit of stampeding is notorious.

This harmonization does not extend to breeding. The Boselaphini have long breeding seasons and produce litters of young—the nilgai produces one or, more commonly, two young, and the Four-horned antelope between one and three. Twins are rare in the Bovini, about one in 40 of calvings in Domestic cattle. These are unwelcome because if one twin is a male and the other a female the latter will usually be sterile; this is because in cattle most twins share a common blood supply and hormones and cells can be exchanged.

The length of the rut may depend to a great extent on the availability of food; year-round sufficiency of food leads to year-round breeding in Domestic cattle. Bison rut in the summer, and births coincide with the spring flush of grass.

A bull detects that a cow is in heat by sniffing her urine and genitals—while doing this, he displays the flehmen, or lip-curl, reaction which, with the pumping action of the tongue, forces scent to the Jacobson's organ above the palate. In all the Bovini the cow becomes attractive to the bull several hours before she ceases to run away whenever he tries to mount her. In commercial herds of cattle, receptive cows mount other cows—in the Chillingham wild cattle this is hardly ever seen because the bulls are efficient at detecting receptive cows; then they do their best to stop the cow from

moving away or associating with other animals. The rutting behavior and fighting of the bull Bovini leads to a brief mating bond with a cow (see pp114–115). If he is not displaced by another bull, he will in due course mount her and inseminate her.

When about to give birth, Domestic cows and the Chillingham cows tend to move away from the herd. African buffalo usually calve within the herd and the cow stays and defends her calf fiercely if the herd should then move on. The gaur is perhaps less efficient at protection—tigers frequently find and kill calves. American bison cows sometimes leave the herd to give birth (lying down to calve); they, too, defend their calves. The Chillingham cows calve standing up; the cow eats the afterbirth. She licks the newborn calf to prevent infestation by flies and to stimulate it to defecate. The calf soon attempts to stand and reach the udder. When the calf has fed, the cow usually rejoins the herd and returns at intervals to feed the calf. After about four days, Chillingham calves, instead of lying down after being fed, follow their mother back to the herd. The other cows show great interest in the new arrival and the mother sometimes chases them away. The bulls are less interested.

In the African buffalo, fertility seems to stay constant as population density increases; in other species, cows may not come into heat when conditions are bad (this is noticeable in the banteng in Java and in the Chillingham wild cattle) but this is a response to the physical rather than to the social environment.

Some of the Bovini were once exceedingly numerous: 40–60 million bison roamed North America. In 1898 the explorer Wellby wrote of Tibet that "on one green hill there were I believe more yak visible than hill." By contrast, some species were probably always rather rare. The gaur, banteng and kouprey need grassy glades in forests, which were probably uncommon before man started practicing shifting agriculture. And crop-raiding aurochsen may have been the first domesticated cattle.

Paradoxically, the success of the Bovini is the cause of the main threat to the survival of some of the species. Any habitat that can support wild cattle can support Domestic cattle, and man's herds carry diseases that can wipe out wild populations. In countries hungry for meat and trophies, big wild hoofed mammals are a great temptation. For these reasons, the survival of most of the wild Bovini can only be assured in properly protected reserves. Indeed, the European bison died out in the wild in 1919 and the present thriving populations in Poland and the Caucasus (USSR) were built up from zoo stocks, then released back into the wild. This may be the only hope for the kouprey. There is also a need to conserve the rare and vanishing breeds of cattle. These, together with the wild species, comprise genetic banks which may well be a valuable source of new varieties and hybrids.   SJGH

The **spiral-horned antelopes** are an African offshoot from the boselaphine lineage but differ from the boselaphines and most African antelopes in apparently lacking territoriality. In this they resemble the Bovini, despite having a very different feeding adaptation, so that comparisons of this tribe with cattle and with other antelopes are likely to give new insights into the evolution of mammalian social organizations. The possibility of domesticating one species, the Common eland, may provide an alternative to cattle in marginal habitats.

More slenderly built than the Bovini and with long, narrow faces, the elegant spiral-horned antelopes are further distinguished by long horns, corkscrewing backwards in the plane of the face. All species are to some extent sexually dimorphic, males tending to be heavier and darker than females and either being the only sex with horns or else having the heavier, more robust horns. They have sharp senses: their greatly enlarged ears suggest that sound is particularly important, but they vocalize rarely. Calls include a few mother/infant calls, an alarm bark and some courtship calls, which often resemble maternal calls.

Even the gregarious, open-country elands are shy; smaller spiral-horned antelopes are mainly active either during the night or at dawn and dusk, and are rarely seen even when at high densities. All species have a white underside to the tail and several have short bushy tails which they turn up sharply when in flight, so showing a conspicuous white "flag."

Between them, the nine species of spiral-horned antelopes have exploited many of the major African habitats south of the Sahara. They are often found either in the zones between vegetation-types or in seasonally variable habitats, where their opportunistic feeding may give them an advantage over more specialized feeders. Their food preferences and their lack of the explosive speed in flight from predators found in many open-habitat species seem to have limited the smaller spiral-horned antelopes to woodlands and other closed

habitats. Even so, the bushbuck, normally found in dense woodland, can use local cover, such as that given by thickets associated with termites' nests in otherwise open grasslands, to exploit a wide range of habitats. The sitatunga lives in marshes, swamps and reed-beds, and the nyala in and around riverside forest and similar habitats in southeastern Africa. The Mountain nyala, only recognized as a species at the beginning of this century, has a limited distribution but is apparently well adapted to its stronghold in the heaths and forests of the Ethiopian highlands. The large, gregarious eland species and, to a certain extent, the kudus, exploit more open woodlands and even grasslands, so following the general trend for large antelope species to be found in more open, arid habitats than their smaller relations. Some nomadic Common eland herds penetrate into extremely dry thorn scrub areas of the Namib and Kalahari deserts, taking advantage of seasonal browse flushes and using water pans as refuges. However, the bongo (the largest known forest antelope) is found in various dense East, central and West African woodlands, a notable and so far unexplained exception to the large species/open habitat rule.

▲ **Wallowing water buffalo** in Sri Lanka. Water buffalo wallow in mud to form a protective cake on their skin which may shield the body from solar radiation as well as providing a protection against biting insects.

▶ **More graceful** than cattle, the Common eland has a long, narrow face and elegantly fluted spiral horns.

▶ **Horn-tangling in Common elands.** Male elands assess each other's status by means of ritual confrontations (horn-tangling). (1) As a preliminary they thresh their horns through aromatic shrubs, collecting gummy saps; (2) they then drive them into the mud, often where another eland has urinated; (3) the horn-tangling encounter itself involves pushing with the horns gently engaged. The smelly mud and sap on the horns advertises prowess at horn-tangling.

All spiral-horned antelopes feed opportunistically off a wide range of food types and species. Their ability to pick out scant high-quality foods from much poorer surrounding vegetation has led to their being described as "foliage gleaners." They may take fruits, seed pods, flowers, leaves (both monocotyledon and dicotyledon species), bark and tubers. Horned species are known to break down otherwise out-of-reach branches and to dig up tubers.

As in most other antelope groups, the smallest, the bushbuck, probably has the highest-quality diet and takes the least grass. The much larger eland species can subsist on much poorer forage but are still much more selective than are similarly sized cattle species, and all body structures related to food-processing reflect this. Spiral-horned antelopes have small, low-crowned teeth, their muzzles are narrow and long, and their digestive systems are not particularly specialized for processing protein-poor, high-fiber food. Cattle species, on the other hand, are mainly grazers and so have to cope with highly fibrous food: they have much more massive teeth, broader muzzles and more elaborate stomachs.

Little is known of the social organization of some spiral-horned antelopes but only one species, the bushbuck, shows a weak form of the territoriality typically found in the boselaphines. Males generally seem to rely on established dominance hierarchies to determine access to females when in large herds, and on dominance displays when strange males meet, in the more solitary species. This convergence with cattle in social organization seems to have had anatomical consequences: the specialized facial and foot glands found in the territorial boselaphines are missing, and body size is emphasized by contrasting markings, by crests and by dewlaps as in the Bovini.

The spiral-horned antelopes tend to have smaller group sizes than do comparable cattle. This may follow from differences in feeding behavior: spiral-horned antelopes often move quite rapidly and erratically through dense woodland while searching for high-quality food. If they were in the large herds typical of the larger Bovini, not only would group members have great trouble coordinating their activities in poor visibility, but also there would be great competition as many animals tried to glean scarce, high-quality food items in the same area.

Females reach sexual maturity at about 18 months (bushbuck) to 3 years (elands). Males in most species continue to grow beyond sexual maturity and it is usually about this age that they start to become noticeably distinct from the females in coloring and build, although their horns will have been growing and differentiating since birth. Males probably breed successfully from 1–3 years after their female peers: they are capable of breeding much earlier, but social factors probably prevent this.

All spiral-horned antelopes almost invariably bear single calves. A given cow probably bears only one calf per year, although the calving rate for bushbuck, which have a gestation period of about five months and may be found in habitats which are only weakly seasonal, may approach two per year. Calving dates vary both with species and with habitat: populations in strongly seasonal habitats (desert fringes, extreme highlands, southern Africa generally) tend to have more sharply defined breeding seasons than those in less variable habitats (riverside forests, equatorial woodlands). In southern Africa, Greater kudu calves are often dropped in the middle or towards the end of the rainy season, while eland cows in the same area will give birth just before the rains begin, a difference of up to six months. Calves wean at from 3–6 months.

Spiral-horned antelopes, particularly the smaller ones, are eaten by a wide range of terrestrial predators. Even the elands, the largest of all antelopes, may be taken by lions or hyenas. The smaller species probably avoid predation mainly by concealment—the sitatunga may actually submerge itself for some time, with just its nostrils above the water, while the larger species will usually flee. Eland cows are known to attack predators near their calves; other spiral-horned antelopes may also do this, but the main defense against predation for newborn calves is to lie concealed in dense undergrowth or tall grass away from the herd for up to a month, being visited by their dams several times a day to nurse. All species have a gruff alarm bark.

When calves join the herd after lying out, they often form a distinct nursery group and have very little to do with adults except for their mothers. Female calves tend to stay with their mothers more than males do, so that most female groups of the smaller species probably consist of related animals (from 2–3 in the bushbuck, up to 10 or 15 in the Greater kudu). Males usually split off from their original group, perhaps as soon as they are weaned, but in any event before they are sexually mature, and form groups of from 2–30 (depending on area and species) males of mixed ages. Older males are more solitary than young ones. The elands differ from this slightly in that small cow groups aggregate during the conception peak into herds of up to several hundreds; many males associate with these herds of females.

▲ **Eland herd.** Unusually for spiral-horned antelopes, elands form herds of up to several hundreds in the breeding season.

◄ **The Greater kudu** shows extreme differences between the sexes, with the females, as here, lacking horns and being considerably smaller than the males.

▼ **The Giant eland** is more of a woodland species than the Common eland, although seen here in wooded savanna.

Regardless of the season, female groups are often accompanied by one or more males, but this is often only a loose association and a given group of females may be accompanied by several different males over a matter of days. During the conception peak, however, a given male may, if not driven off, escort a female for several days. Calves are therefore probably fathered by the most dominant bull around when cows come into heat.

Most spiral-horned antelopes, particularly the elands and kudus, are highly mobile and probably have large home ranges compared with similarly sized antelopes. Most species are resident in a given area throughout the year, the ranges of adjacent groups overlapping considerably. The elands travel large distances through the year and some populations must be regarded as migratory, although they do not necessarily form large, conspicuous herds in their yearly moves from, for example, highland plateau grasslands to lowland savannas. Other species show a more subtle shift in habitat use: Greater kudus may move up and down hillsides and vary their use of areas around the hills to take advantage of seasonal changes in the vegetation that they consume.

All the species of spiral-horned antelope are fairly secure, at least in the short term, even the Mountain nyala, once thought to be highly endangered. Various populations within species are, however, less safe. In particular, the low-density, highly mobile elands are vulnerable to any agricultural development and to hunting. The refuges provided by nature and forestry reserves, and by some game ranching efforts, are particularly important, since the spiral-horned antelopes have largely disappeared from areas cleared for agriculture. The precarious state of the Western race of the Giant eland, which was probably originally a scattered series of small local populations, should warn against allowing other species to be whittled back to a few small strongholds.

Several species are known to have suffered badly from rinderpest. The various ranching experiments with elands and kudus also suggest that the spiral-horned antelopes are particularly vulnerable to heavy tick infestations and to resulting tick-borne diseases. Bushbuck and kudus may act as reservoirs of sleeping sickness in tsetse infested areas.

Perhaps because of their combination of size and elegance, the larger spiral-horned antelopes, particularly the elands, have a special place in African tradition. Rock paintings in the Kalahari include sketches of Common eland apparently in some form of domestic relationship with the local bushmen. In more recent times, herds of Common eland have been kept under various regimes as dairy animals (Askanya Nova, USSR), draft animals (Natal, South Africa), and as either a direct substitute for more conventional livestock in marginal habitats or else, as with many other species of hoofed mammals, as part of a game-cropping scheme. The success of such efforts seems to depend very much on local conditions and the individuals involved: large-scale plans for land use, now being prepared in several African countries, should provide more and better opportunities for such development. Experiments in domestication are doubly important since similar efforts may have already saved the blesbok and the Black wildebeest from extinction.

THE 23 SPECIES OF CATTLE AND SPIRAL-HORNED ANTELOPES

Tribe Boselaphini

Both species in peninsular India. Relics of the stock from which the Bovini arose. They have primitive skeletal, dental and behavioral characteristics; females not horned.

Genus *Tetracerus*

Four-horned antelope [*]

Tetracerus quadricornis
Four-horned antelope or chousingha.

Wooded, hilly country near water. HBL 39in; TL 5in; HT 24in; HL (male) front pair ¾–1½in, rear 3–4in; WT 44lb. Coat: male dull red-brown above, white below, dark stripe down front of each leg; old males yellowish; females brownish bay. Facial and foot glands.

Genus *Boselaphus*

Nilgai

Boselaphus tragocamelus
Nilgai or Blue bull.

Thinly wooded country. Rather horse-like. HBL 79in; TL 18–20in; HT 47–59in; HL (male) 6–9in; WT male 530lb, female 265lb. Coat: coarse iron-gray; white markings on fetlocks, cheeks, ears; tuft of stiff black hairs on throat; young bulls and cows tawny.

Tribe Bovini
(Wild Cattle)

N America to Mexico; Africa, Europe, Asia, Philippines and Indonesia. Introduced to Australia and New Zealand. Stout bodies, low, wide skulls and smooth or keeled horns in both sexes, splaying out sideways from skull. No facial or foot glands. Sexual maturation and frequency of calving depend to a degree on health and feeding, but all those which have been studied can achieve maturity at 2 years and produce a calf a year for most of their life. Males may also be able to fertilize at 2 years but are usually prevented by social factors. Longevity around 15–20 years; this, and most of the body dimensions quoted may seldom if ever be achieved in most wild populations now.

Genus *Bubalus*
(Asiatic buffalos)
Stout, dark-colored bodies, hair generally sparse. Horn cores triangular in cross section. No boss of horn in either sex; no dewlaps or hump.

Wild water buffalo [V] [*]

Bubalus arnee (bubalis)
Wild water buffalo, carabao or arni.

Very widespread (Asia, S America, Europe, N Africa) as domestic form, of which there are two groups of breeds, the River and the Swamp buffaloes. Latter is feral in N Australia. The wild race survives as remnants in reserves or in remote places in India, Assam, Nepal, Burma, Indochina and Malaysia. Near (and in) large rivers in grass jungles and marshes. HBL 94–110in; TL 24–33in; HT 63–95in; WT male 2,650lb, female 1,765lb. Coat: slaty black, legs dirty white up to hocks and knees; white chevron below jaw. Flexible fetlock joints make it nimble in the mud; horns heavy, backswept. Gestation 310–330 days.

Lowland anoa [E]

Bubalus depressicornis

Confined to swampy lowland forests of N Sulawesi (Celebes). A miniature Water buffalo—almost deer-like. Biology almost unknown. HBL about 71in; TL about 16in; HT about 34in; HL male 12in, female 10in. Coat: black, sparse hair and white stockings; young woolly, brown. Horns flat and wrinkled.

Mountain anoa [E]

Bubalus quarlesi

Forest at high altitude in Sulawesi. Adults look like juvenile Lowland anoa. Biology almost unknown. HBL about 60in; TL about 10in; HT about 27in; HL 6–9in. Coat: legs same color as body which is brown-black; woolliness persisting into adulthood. Horns smooth and circular.

Tamarau [E]

Bubalus mindorensis
Tamarau or tamaraw.

Found only in forest on the island of Mindoro (Philippines). Biology unknown; said to be nocturnal and ferocious. HT 39in; HL about 14–20in. Coat: dark brown-grayish black. Horns stout. Gestation 276–315 days.

Genus *Bos*
(True cattle)
Dark short-haired coat (excepting yak and some Domestic cattle). Horn cores circular in cross section. No boss of horn in either sex; dewlaps and humps well developed in some species.

Banteng [V]

Bos javanicus
Banteng, tsaine, tembadau.

Isolated populations in Indochina and on the islands of Borneo, Java and Bali (domesticated as the Bali cattle). Feral in N Australia. Quite thick forest with glades. Very like Domestic cow in general proportions. HBL (mainland race) 75–88in; TL 25–28in; HT 63in; WT 1,325–1,765lb. Coat: adult bulls dark chestnut; young bulls and cows reddish brown; all have white band around muzzle, white patch over eyelids, white stockings and white rump patch; males have horny bald patch of skin between horns, and a dorsal ridge and dewlap. Gestation about 285 days.

Gaur [V]

Bos gaurus
Gaur, Indian bison or seladang.

Scattered herds in peninsular India, a few surviving in Burma, W Malaysia and elsewhere in Indochina. Upland tropical forest with glades. HBL 98–120in; TL 27–39in; HT 67–79in; HL (male) up to 31in; WT male 2,075lb, female 1,545lb. Coat: adult bulls shiny black with white stockings and gray boss between the horns; young bulls and cows dark brown, also with white stockings. Huge head, deep massive body and sturdy limbs. Hump, formed by long extensions of the vertebrae, small dewlap below the chin and a large one draped between the forelegs; the horns sweep sideways and upwards. Gestation: 9 months.

Yak [E]

Bos mutus (grunniens)

Scattered localities on alpine tundra and ice deserts of the Tibetan Plateau at altitudes of 13,130–19,700ft. Biology almost unknown, though domesticated yak have been studied. A 1938 report records a bull as 80in high at the shoulder, with horns of 31in and weighing 1,810lb; a cow as 61in high, with horns of 20in and weighing 675lb. Coat: shaggy fringes of coarse hair about body with dense undercoat of soft hair; blackish brown with white around the muzzle. Massively built with drooping head, high humped shoulders, straight back and short sturdy limbs. Gestation: 258 days.

Cattle [D]

Bos primigenius (taurus)

By convention, present-day cattle are given the same scientific name as that of their ancestor, the aurochs or urus which died out in 1627. All cattle are completely interfertile but the skull and blood proteins of the humped Zebu cattle are different from those of the humpless breeds; perhaps the domestications were of two races of the aurochs. The hump of the Zebu, unlike those of the other species mentioned in this table, is composed of muscle and fat and is not simply the result of long processes on the vertebrae. Domestic cattle are feral in many places, the most notable being Chillingham Park, N England. Cattle exist in many sizes and forms. There are long-horned and polled (hornless) breeds; generally shoulder height is 71–79in and weight 990–1,985lb. Gestation: around 283 days.

Kouprey [E]

Bos sauveli
Kouprey or Cambodian forest ox.

A few living in forest glades and wooded savannas of Indochina. Discovered in 1937. HBL 83–87in; TL 39–43in; HT 67–75in; WT 1,540–1,985lb. Coat: old bulls black or very dark brown; may have grayish patches on body; cows and young bulls gray, underparts lighter and chest and forelegs darker; both sexes have white stockings. Bulls have very long (over 16in) dewlap hanging from neck; dorsal ridge not well developed; horns in female are lyre shaped; in males horns curve forwards and round, then up; horn tips are frayed.

Genus *Synceros*
(African buffalos)
Africa south of the Sahara. Bulky, dark. Horn cores triangular in cross section. Males have heavy boss of horn on top of head; this, with the elaborate submissive display differentiate the genus from *Bubalus*.

African buffalo

Synceros caffer

Generally considered as two subspecies: the Cape buffalo (*S.c. caffer*), and the Forest buffalo (*S.c. nanus*), with intermediate forms. The former lives in savannas and woodlands, the latter in forests nearer the Equator. The Cape buffalo has HBL 94–135in; TL 30–43in; HT 53–70in; HL 20–60in; WT male 1,500lb, female 1,060lb. Coat: brownish black. The Forest buffalo has HBL 87in; TL 27in; HT 39–47in; WT male 106–126lb, female 585lb. Coat: reddish brown. Gestation: 340 days.

Genus *Bison*
(American and European bisons)
Reddish brown–dark brown coat;
long shaggy hairs on neck and head,
shoulders and forelegs, and a beard.
Short and broad skull, dorsal hump
formed by processes of the vertebrae.
Smooth horns, circular in cross
section, about 18in. These species
are completely interfertile.

American bison ⊡
Bison bison
American bison or buffalo.

Widely considered to be two
subspecies, the Plains bison (*B.b.
bison*) and the rather larger and
darker Wood bison (*B.b. athabascae*),
which lives further north. N America.
Grassland, aspen parkland and
coniferous forests. Associated now
with prairies but inhabited forests as
well before its virtual extermination
last century. Now mainly in parks
and refuges. HBL 150in; TL 35in; ht
male 77in; wt male 1,803lb, female
1,201lb. Gestation: 270–300 days.

European bison
Bison bonasus
European bison or wisent.

Became extinct in the wild in 1919
(Russian-Polish border) but was re-
established in Bialowieza Primeval
Forest and later in the Caucasus and
elsewhere in the USSR. Mixed woods
with undergrowth and open spaces.
HBL 114in; TL 31in; ht 71–77in;
wt male 315lb. Gestation:
254–272 days.

Tribe Strepciserotini
(Spiral-horned antelopes)
Restricted to Africa. Medium to large
body size, more slenderly built than
the Bovini, with long necks and deep
bodies. Adult males larger than
females; sexes also differ in markings
and horn structure. Horns in males
only in most species. No distinct facial
or foot glands.

Genus *Tragelaphus*
(Kudus and nyalas)

Sitatunga
Tragelaphus spekei
Swamps, reedbeds and marshes of the
Victoria, Congo and Zambezi-
Okavango river systems. Male HBL
59–67in; TL 8–10in; HL 18–35in;
wt 176–275lb. Female HBL 53–61in;

TL 8–10in; wt 110–132lb.
Coat: shaggy and slightly oily;
female lighter and redder in color
than the yellowish to dark gray-
brown males; spots and up to ten
white stripes on the body; dorsal crest
runs the length of the body and is
erectile; white patches on throat,
white spots on cheeks. Long splayed
hooves distribute weight, allowing the
animal to walk through mud without
sinking into the ground.

Nyala
Tragelaphus angasi
Riverside thicket and dense
vegetation in SE Africa. Male HBL
83in; TL 17in; SH 44in; HL 25in;
wt 236lb. Female HBL 70in; TL 14in;
SH 38in; wt 137lb. Coat: shaggy,
dark gray-brown, particularly
along the underside of the
body and throat; usually several
poorly marked white vertical stripes;
long, conspicuous erectile crest,
brown on the neck and white along
the back; legs orange; white chevron
between the eyes; horns lyre-shaped
with a single complete turn, black
with whitish tips; females much
redder with clearly marked white
stripes, short coats, a less obvious
chevron, and generally resemble
bushbuck females.

Bushbuck
Tragelaphus scriptus
Locally throughout Africa south of
the Sahara, except for the arid
southwestern and northwestern
regions, in a wide range of habitats
whose common feature is dense
cover. Male HBL 45–57in; TL 8–9in;
HL 10–22in; wt 66–165lb.
Female HBL 43–51in; TL 8–9in;
wt 53–92lb. Coat: short, varying
from bright chestnut to dark brown,
with white transverse and vertical
body stripes being either clearly
marked, broken, or reduced to
a few spots on the haunches; black
band from between the eyes to the
muzzle; white spot on cheek, two
white patches on the throat; adult
males are darker than females and
young, especially on the forequarters,
and the erectile crest is more
prominent.

Mountain nyala
Tragelaphus buxtoni
Highland forest and heathland of the
Arusi and Bale Mountains in Ethiopia.
HBL 75–98in; HL up to 12in.
Coat: shaggy, grayish brown; about
four ill-marked vertical white stripes;
white chevron between the eyes; two
white spots on the cheeks; two white
patches on the neck; short white
mane continued as a brown and
white crest. Horns one to one-and-a-
half fairly open turns. In many ways
resembles the Greater kudu more
closely than the nyala.

Lesser kudu
Tragelaphus imberbis
Thicket vegetation in Ethiopia,
Uganda, Sudan, Somalia, Kenya and
N and C Tanzania. HBL 63–69in;
TL 10–12in; HL 24–35in; wt (male)
200–242lb, female 121–154lb. Coat:
sleek and short haired, brownish-gray
with 11–15 clearly marked vertical
white stripes; head darker with
incomplete white chevron between
the eyes; two white patches on the
neck; male's dorsal crest extends
forwards into a short mane; tail
bushy; small but clear spots on
cheeks; reddish tinge to legs; female
slightly more reddish than male.
Horns 2–3 open spirals.

Greater kudu
Tragelaphus strepsiceros
Woodland, especially in hilly, broken
ground, in E, C and S Africa. Male
HBL 75–99in; TL 14–19in;
HL 39–71in; wt 420–695lb. Female
HBL 75–87in; TL 14–19in; wt
265–475lb. Coat: short, blue-gray to
reddish brown; 6–10 vertical white
body stripes; white chevron between
eyes; up to three white cheek spots;
dorsal crest extended by mane along
whole body; fringe of hairs from chin
to base of neck in males; females and
young redder than males.

Bongo
Tragelaphus euryceros
Discontinuously distributed in
lowland forest in E, C and W Africa;
found outside this habitat in S Sudan,
in small populations in montane or
highland forest in Kenya, and in the
Congo. Male HBL 87–92in; TL 10–11in;
HL 24–39in; wt 530–895lb.
Female HBL 87–92in; TL 9–10in;
HL 24–39in; wt 460–560lb.
Coat: bright chestnut red,
much darker in adult males; dark

muzzle, white chevron between eyes,
about 2 white cheek spots; lower neck
and undersides darker, whitish
crescent collar at base of neck; black
and white spinal crest and many
narrow but clear white vertical stripes
on the body; contrasting black and
white markings on the legs. Horns
present in both sexes, heavy and
smooth with an open spiral of one to
one-and-a-half turns. Tail long and
tufted at tip.

Genus *Taurotragus*
(Elands)

Common eland
Taurotragus oryx
Common or Cape eland.
Nomadic grassland and open
woodland; may have at least visited
all but the most arid of these habitats
in E, S and C Africa in the past; now
found only in game reserves and
ranches in some areas. Male HBL
98–134in; TL 21–30in; SH 53–70in;
HL 24–40in; wt 880–2,100lb.
Female HBL 79–110in; TL 21–30in;
SH 49–59in; HL 24–55in;
wt 860–1,310lb. Coat: light
tan, darkening to gray in old
males; a few light stripes on the
forequarters (not found in adults of
the southern populations); black and
white leg markings; black tuft on end
of tail; black stripe along back,
merging into short mane; adult males
develop a tuft of frizzy hair on their
foreheads. Ears have much smaller,
more horse-like pinnae than other
spiral-horned antelopes.

Giant eland
Taurotragus derbianus
More of a woodland species than the
Common eland and found in small
populations in W and C Africa, with
larger and more secure populations
in E Africa, particularly Sudan. Male
HBL 114in; ht 59–69in;
TL 22–31in; HL 31–43in;
wt 990–1,990lb. Female HBL 87in;
ht 59in; TL 22–31in; HL 31–49in;
wt 970lb. Coat: reddish brown,
becoming slate gray in adult males;
12–15 body stripes, white chevron
between eyes, white cheek spots,
black stripe along back merging into a
short mane, black collar around neck,
with contrasting white patches on
either side. The collar emphasizes the
dewlap, which begins under the chin
and finishes above the base of the
neck. Horns more strongly keeled
than in the Common eland, but more
slender and longer, even in the males.

Bison Breeding

Mating system of American bison

An all-out fight between American bison bulls is one of the great dramas in nature. They slam their heads together, their hooves churning up dust in enveloping clouds. Clumps of hair two or three times the size of a man's fist are sheared from their heads by their horns grinding against each other, and tossed into the air. They circle, trying to exploit the agility conferred by their small hindquarters, to drive a horn into the opponent's ribcage or flank. If one succeeds, the other may die of the wound.

Fighting is as costly as it is spectacular: it costs time and energy, and the combatants risk injury, even death. In general, costly behavior is rare. Fewer than 15 percent of bull bison disputes are settled by fights. All the rest are settled by signals of threat and yielding, of several degrees of intensity and sometimes surprising subtlety. A bull may threaten by standing broadside to his opponent, by bellowing, by rolling in dust (sometimes he urinates in the dust before rolling in it) and by approaching his opponent head on. Sometimes two bulls stand close together, heads to the side, and "not threaten." swinging their heads up and down in matched movements. To yield, a bull may simply withdraw, at other times he may duck his head and turn it away from his opponent. Sometimes he grazes rather ceremoniously.

An observer knowing these signals can recognize which bull of a pair is dominant. Dominance relations between bison bulls often change after only a few days, but they are nevertheless important. During the two weeks of each year when 90 percent of the breeding takes place, the dominant bull of any pair can displace the subordinant from a receptive cow. The more other bulls a particular bull dominates, the more cows to which he has priority. As a result, the one-third of the bulls that are the most dominant mate with about two-thirds of the cows each year. A few bulls serve as many as 10 cows in a season.

This variation in male reproductive success helps to explain the great differences in the bodies of bulls and cows. The bulls give no care to their young. They are specialized for breeding, and in bison society that means specialized for threatening and fighting. In 1871, Darwin pointed out that selection for such breeding characteristics is a special case of natural selection, which he named "sexual selection." The bull bison is an excellent example of a fighting specialist: male reproductive success is determined by competition for females; female reproductive success varies much less, and is deter-

mined by competition for food, not mates.

Bison mate in temporary one-to-one relationships called tending. The tending bull stands beside a particular cow, threatening bulls that approach. He blocks the cow's path if she starts to move away, and he tries to mount her. This relationship may last for only a few minutes before the bull is displaced by a more dominant bull, or leaves voluntarily to seek a more receptive cow. Few cows or bulls show any preference for particular partners.

Each sex follows its own strategy to maximize reproductive success. The cow has no more than one calf a year. So her strategy emphasizes the quality of the bull with whom she mates. Since more dominant males are likely to have more dominant sons, the female behaves so as to maximize the chance of being mated by a more dominant bull. For several hours

▲ **Classic confrontation.** Bison bulls locked in combat.

▶ **Stages of combat.** Bison bulls use a repertoire of threats in establishing dominance relations, and conflict rarely escalates to all-out fighting. In the broadside threat (1) the animals stand broadside to each other, either facing the same way or in opposite directions. They often bellow and try to present the most impressive profile, arching their backs and lifting their bellies. In the nod-threat (2) the animals stand facing each other and swing their heads from side-to-side in unison. A bull may signal submission during the threat interactions by various means: by turning the head away, by retreating, by lowering the head to graze etc. If no submissive signal is given fighting may begin (3).

istically and unselectively. After mating, a bull usually stays with the cow for half an hour or so, guarding her against the approach of other bulls, and making sure that it is his sperm that fertilizes her. The few cows that copulate more than once a year copulate the second time within an hour of the first, usually with the same bull. Therefore, while bison behave promiscuously in the sense that any cow may mate with any bull, the genetic outcome of their mating pattern is like that of a polygynous mating system, in which a dominant male prevents access by other males to the females in his harem.

Knowledge of their mating system is essential for the development of a proper conservation program for bison. In their natural state, bison lived in large, interacting populations. Genes were exchanged over great distances among millions of individuals. There was little chance of the inbreeding, or gene loss through random failure of some genes to be transmitted (genetic drift), that can occur in small, isolated populations. But today bison, like more and more wild animals, live in reserves, where their populations are often small and isolated. Inbreeding and gene loss are now quite possible, and both are harmful. Whether or not either will occur depends largely on the effective size of the breeding population, which in turn depends more on the breeding system than on the number of males and females of breeding age in the population. For example, if a population of ten males and ten females breeds monogamously, the effective breeding population is 20. But if they breed polygynously, with one dominant male mating with all the females, the effective breeding population is only 3.7. Knowledge of the social system, including the breeding system, of bison will help to determine how large populations must be to survive genetically in the long term. In this way it will make a major contribution to the long-term conservation of this no longer plentiful, but still magnificent, wild animal.   DFL

before she copulates, the female breaks away from tending bulls and runs through the herd, drawing the attention of all the bulls present, and inciting competition between them.

In contrast, bulls may sire several calves each year, so their strategy emphasizes quantity rather than quality. They move quickly from cow to cow, mating opportun-

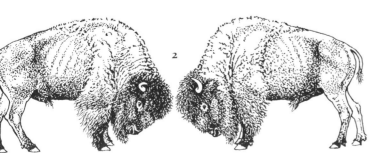

DUIKERS

Subfamily: Cephalophinae
Seventeen species in 2 genera.
Family: Bovidae.
Order: Artiodactyla.
Distribution: Africa south of Sahara.

Habitat: mainly dense forest thickets.

Size: ranges from head-body length
22–29in (55–72cm), tail length
3–5in (7–12cm) and weight
8.8–13.2lb (4–6kg) in the Blue
duiker to head-body length
45–57in (115–145cm), tail length
4.3–7in (11–18cm) and weight
100–176lb (45–80kg) in the
Yellow-backed duiker.

Gestation: 7½–8 months in Red-flanked and Bay
duikers.
Longevity: unknown in the wild; 10–15 years
in captivity.

Forest duikers
Sixteen species in genus *Cephalophus*. Species
include: **Blue duiker** (*C. monticola*); **Yellow-
backed duiker** (*C. sylvicultor*); **Bay duiker**
(*C. dorsalis*); **Maxwell's duiker** (*C. maxwelli*);
Zebra duiker (*C. zebra*); **Jentink's duiker**
(*C. jentinki*).

Savanna or bush duikers
One species: Common duiker (*Sylvicapra
grimmia*).

ALMOST anywhere in Africa, south of the Sahara desert, the observant traveler may catch a glimpse of one of many small to medium-sized antelopes disappearing into dense forest or thickets. These are the "duikers," an Afrikaans word meaning divers, so named for their habit of diving into cover when disturbed. The short forelegs, longer hindlegs, and arched body shape, enable duikers to slip easily through dense vegetation.

The sexes are similar in appearance, although females are often up to 4 percent longer than males. Both usually possess short, conical horns, but these are occasionally lacking in females. The coat is often reddish, but in some species it is blue-gray, black or striped. Duikers have the largest brains relative to body size of all antelopes.

Most duikers are forest duikers of the genus *Cephalophus*. The name *Cephalophus* refers to the crest of long hair between the horns. The genus *Sylvicapra* includes only one species, the Common duiker, which is the only species typically found in the savanna and open bush country.

Duikers are uncommon as fossils, but the living species retain many features found in the fossil remains of early bovids and are sometimes considered to be the most primitive living African antelopes.

Little is known about the natural history of most duiker species in the wild due to the impenetrability of their natural habitat and their secretive nature. The Yellow-backed duiker may be active during both night and day but some species, such as the Blue duiker, are active only during the day, while others, such as the Bay duiker, are active only at night. Duikers are primarily browsers and require high-quality food because of their relatively small size. They feed on leaves, fruits, shoots, buds, seeds and bark. Surprisingly, they sometimes stalk and capture small birds and rodents and occasionally also eat insects and even carrion.

Duikers are not gregarious and are usually seen alone or in pairs. In the wild the Blue duiker is truly monogamous; pairs seem to mate for life and reside in small— 5–10 acre (2–4ha)—stable territories which are actively defended against other members of the same species by both male and female. In captivity, male Maxwell's duikers are most aggressive to unfamiliar males while females are more aggressive towards unfamiliar females. Observations on captive animals suggest that several other duiker species may also be monogamous.

In monogamous social systems, which are comparatively rare in mammals, males sometimes provide a great deal of care for the young, and the young remain in the social group for extended periods during which they often help to care for their younger siblings (obligate monogamy). However, in other monogamous species, males leave most responsibility for the care of the young to the female and young disperse from the social group before the birth of their younger siblings (facultative monogamy). In species exhibiting obligate monogamy, such as marmosets and tamarins, mates remain near one another

▼ **Representative species of duiker.** (**1**) Male Common duiker (*Sylvicapra grimmia*) "duiking" (fleeing). (**2**) Yellow-backed duiker (*Cephalophus sylvicultor*), showing the erect yellow rump patch. (**3**) Male Zebra duiker (*Cephalophus zebra*) marking a sapling with the preorbital gland. (**4**) Jentink's duiker (*Cephalophus jentinki*) suckling a calf. (**5**) Male Maxwell's duiker (*Cephalophus maxwelli*) about to rub his preorbital gland on a rival before fighting. (**6**) Red-flanked duiker (*Cephalophus rufilatus*) eating a bird.

and usually feed and rest together. Blue duikers (along with species in other groups, such as dikdiks and elephant shrews) exhibit facultative monogamy, in which mates do not remain near each other and often feed and rest at different times.

Duikers have only a single calf. Calves of all species spend little time with their mothers during the first few weeks of life, remaining well hidden in vegetation. Young Blue duikers reach sexual maturity at about one year of age and leave their parents during the second year of life to attempt to find their own mates and territories. Larger species probably reach sexual maturity at an older age. Calves of some species, such as the Zebra duiker and Maxwell's duikers, resemble their parents in color while those of other species have distinctive juvenile coats. Young Jentink's and Bay duikers are very similar in appearance, with a uniform dark brown coat unlike that of the adults of either species. The yellow rump patch of the Yellow-backed duiker does not begin to appear until about one month of age and is not fully developed until about 10 months of age.

Duikers possess large scent glands beneath each eye. The structure of these glands differs from that of the preorbital glands found in other antelopes. The secretion, which may be clear or bluish in some species, is extruded through a series of pores instead of a single large opening. In captivity, many duiker species rub these glands on fences, trees and other objects in their enclosures. This has usually been interpreted as territorial marking. This behavior is very frequent in some species: in captivity, male Maxwell's duikers may mark as often as six times during a 10-minute interval.

Maxwell's duikers also press these glands on the glands of other individuals, first on one side and then on the other. This behavior is called mutual marking. Very forceful mutual marking has been observed in captivity between males as a prelude to fighting. A much more gentle form of mutual marking is often seen between males and females.

The rarest species is Jentink's duiker, which is considered to be endangered. It was not described until 1892 and even today few specimens exist in museums and zoos. The main threat to this species, as well as to other duiker species, is subsistence hunting. Duiker meat is well liked by humans and duikers are easily dazzled at night with lights, making them comparatively easy to shoot or capture. Some species are also captured by driving them into nets.   KR/KK

THE 17 SPECIES OF DUIKERS

Abbreviations: HBL = head-body length. TL = tail length. wt = weight.
[E] Endangered. [*] CITES listed.

Genus *Cephalophus*
Forest Duikers
Horns usually in both sexes and in same plane as forehead. Ears short and rounded.

Maxwell's duiker
Cephalophus maxwelli

Nigeria west to Gambia and Senegal. Lowland forest and adjacent gallery forest. HBL 22–35in; TL 3–4in; wt 17–20lb. Coat: from gray-brown to blue-gray, with much variation. Horns sometimes absent in females.

Blue duiker [*]
Cephalophus monticola

Nigeria to Gabon, east to Kenya and south from this latitude to S Africa. Lowland forest. HBL 22–28in; TL 3–5in; wt 9–13lb. Coat: color variable; similar to Maxwell's duiker but may be bluer.

Black-fronted duiker
Cephalophus nigrifrons

Cameroun to Angola and east through Zaire to Kenya. Lowland, gallery, montane and marshy forests. HBL 33–42in; TL 4–6in; wt 28–35lb. Coat: reddish to dark brown, with reddish brown to black stripe from nose to horns.

Red-flanked duiker
Cephalophus rufilatus

Senegal to Cameroun east to Sudan and Uganda. Gallery forests and forest edges. HBL 24–27in; TL 3–4in; wt 20–26lb. Coat: reddish yellow to reddish brown with dark dorsal stripe from nose to tail. Horns regularly lacking in female.

Jentink's duiker [E]
Cephalophus jentinki

Liberia and W Ivory Coast Lowland forest only. HBL about 53in; TL about 6in; wt up to 155lb. Coat: head and neck black, shoulders white, back and rump a grizzled "salt and pepper." A very rare species.

Yellow-backed duiker
Cephalophus sylvicultor

Guinea-Bissau east to Sudan and Uganda, south to Angola and Zambia. Wide variety of forest types and open bush. HBL 45–57in; TL 4–7in; wt 99–176lb. Coat: blackish brown except for yellow rump patch..

Abbott's duiker
Cephalophus spadix

Tanzania. High montane forest. HBL 39–47in; TL 3–5in; wt up to 132lb. Coat: dark chestnut brown to black.

Zebra duiker
Cephalophus zebra

Sierra Leone, Liberia and Ivory Coast. Lowland forest. HBL 33–35in; TL about 6in; wt 20–35lb. Coat: reddish brown with 12–15 black transverse stripes.

Black duiker
Cephalophus niger

Guinea east to Nigeria. Lowland. forests. HBL 31–35in; TL 5–6in; wt 33–44lb. Coat: brownish black to black.

Ader's duiker
Cephalophus adersi

Zanzibar, coastal Kenya and Tanzania. HBL 26–28in; TL 3–5in; wt 14–26lb. Coat: tawny-red with white band on rump.

Red forest duiker
Cephalophus natalensis

Somalia south to Zimbabwe and Mozambique. Lowland montane forest. HBL 27–39in; TL 4–6in; wt 23–26lb. Coat: orange -red to dark brown.

Peter's duiker
Cephalophus callipygus

Cameroun and Gabon east through Central African Republic and Zaire. Lowland forest. HBL 31–45in; TL 4–7in; wt 33–53lb. Coat: light to dark reddish brown.

Cephalophus weynsi

Zaire, Uganda, Rwanda and W Kenya. A very poorly known species which may be a distinct species or a form of either Peter's duiker or the Red forest duiker.

Bay duiker
Cephalophus dorsalis

Guinea-Bissau east to Zaire and south to Angola. Lowland forest. HBL 27–39in; TL 3–6in; wt 42–55lb. Coat: brownish yellow to brownish red; dorsal black stripe from nose to tail.

White-bellied duiker
Cephalophus leucogaster

Cameroun south and east into Zaire. Lowland forest. HBL 35–39in; TL 5–6in; wt 26–33lb. Coat: light to dark reddish brown with white chin, throat, and belly.

Ogilby's duiker
Cephalophus ogilbyi

Sierra Leone east to Cameroun and Gabon. Lowland forests. HBL 33–45in; TL 5–6in; wt 31–44lb. Coat: reddish orange with white stockings; dorsal black stripe from shoulder to tail.

Genus *Sylvicapra*
Savanna or Bush duikers
Horns usually in males only and directed upward. Ears longer and more pointed than in *Cephalophus*.

Common duiker
Sylvicapra grimmia

Subsaharan Africa except Zaire. Savanna and open bush. HBL 31–45in; TL 4–9in; wt 22–40lb. Coat: sandy-tan.

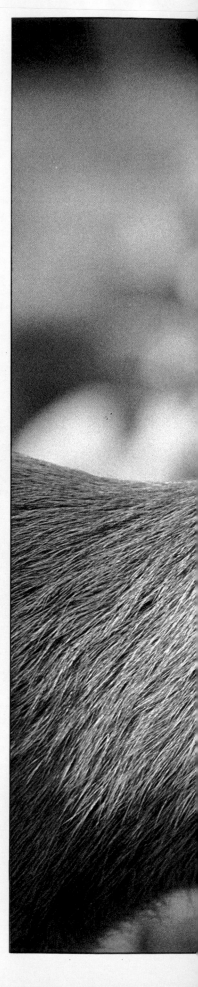

► **A Common duiker** in the Winterberg Mountains, Nambia

GRAZING ANTELOPES

Subfamily: Hippotraginae
Twenty-four species in 11 genera.
Family: Bovidae.
Order: Artiodactyla.
Distribution: Africa, Arabia.

Habitat: dry and wet grasslands up to 16,400ft
(5,000m).

Size: shoulder height from 26–30in (65–76cm)
in the Mountain reedbuck to 49–57in
(126–145cm) in the Roan antelope; weight
from 51lb (23kg) in the Grey rhebok to
620lb (280kg) in the Roan antelope.

Reedbuck, waterbuck and rhebok
Tribe: Reduncini.
Nine species in 3 genera.
Species include: **Bohor reedbuck** (*Redunca
redunca*), **Gray rhebok** (*Pelea capreolus*), **kob**
(*Kobus kob*), **lechwe** (*Kobus leche*), **Mountain
reedbuck** (*Redunca fulvorufula*), **Nile lechwe**
(*Kobus megaceros*), **puku** (*Kobus vardoni*),
waterbuck (*Kobus ellipsiprymnus*).

Gnus, hartebeest and impala
Tribe: Alcephalini.
Eight species in 5 genera.
Species include: **bontebok** (*Damaliscus dorcas*),
Brindled gnu (*Connochaetes taurinus*),
Hartebeest (*Alcephalus busephalus*), **hirola**
(*Beatragus hunteri*), **impala** (*Aepyceros
melampus*), **topi** (*Damaliscus lunatus*), **White-
tailed gnu** (*Connochaetes gnou*).

Horse-like antelope
Tribe: Hippotragini.
Seven species in 3 genera.
Species include: **addax** (*Addax nasomaculatus*),
Arabian oryx (*Oryx leucoryx*), **bluebuck**
(*Hippotragus leucophaeus*), **gemsbok** (*Oryx
gazella*), **Roan antelope** (*Hippotragus equinus*),
Sable antelope (*Hippotragus niger*), **Scimitar
oryx** (*Oryx dammah*).

▷ **Apparently marooned,** these waterbuck
are browsing on a patch of aquatic vegetation.
They are never far from water, which they need
in quantity to accompany their high-protein
diet.

▶ **The Bohor reedbuck,** an antelope of the
northern savannas of Senegal, east to Sudan
and south across Tanzania.

THE grasses of Africa lie in an unbroken
prairie which stretches from the cold
subdesert steppes of the Cape Province of
South Africa, through the deciduous woodlands and open grasslands of the southern
savannas, crosses the equator in East Africa,
and spreads out across the northern savannas, reaching up to the Sahara desert. Even
in the heart of the desert, the soils can throw
up a rich growth of grass following the
passage of a rain storm. Interspersed with
the prairie grasses are the papyrus-filled
lakes and swamps, marshes, water
meadows, reed-beds, and the grass-bound
flats and floodplains of Africa's great river
systems, notably the Nile, Niger, Zaire and
Zambesi. Rising above these wetlands, the
Adamawa highlands of Cameroon, the
highland massifs of Ethiopia and Kenya, and
the Drakensberg Mountains of South Africa
and Lesotho are clad in montane grasslands.
Africa's diverse grasslands are home to the
grazing antelopes (subfamily Hippotraginae), a group which have colonized every
habitat from the inundated swamps of the
Nile Sudd to the barren centers of the Namib
and Sahara Deserts and the exposed mountain pastures up to 16,400ft (5,000m) on
Mt. Kilimanjaro.

The wetlands and, rather unexpectedly,
the montane grasslands are inhabited by the
reedbucks, kobs and waterbuck, all from the
tribe Reduncini. The smaller reedbucks
have retained many primitive features, and
the Mountain reedbuck particularly may be

taken as an approximate model, in general
appearance and behavior, of the ancestors
of this group. Indeed, the existence of three
highland populations, each separated by
1,240mi (2,000km) from its nearest neigh-
bor, suggests that a single dominant
ancestral stock was once widespread in all
types of grassland, subsequently relinquish-
ing the lowlands to more advanced forms.
Adapted to a poor-quality, fibrous diet, the
diminutive Mountain reedbuck is particularly sedentary. The females and young
either live within the territories of single
resident males or range over a small number
of neighboring territories, normally in
groups of 2–6. The average size of territory
varies from 25–37 acres (10–15ha) in one
Kenyan population of moderately high density (28 animals per sq mi; 11 per sq km).
These territories are not marked by glandular secretions, dung piles or scrapes; the
principal advertisement of presence is the
animals' whistle. The population structure
of the Gray rhebok, another small montane
antelope, is very similar.

The Southern reedbuck and Bohor reedbuck are lowland species which occur in the
southern and northern savannas respectively. Seldom far from water, they are
typical inhabitants of floodplains and
inundated grasslands. They are particularly
active at night, emerging from dense cover
to feed on open lawns, to the accompani-
ment of much whistling and bouncing. In
farming areas, young cereal crops are a

A High-density Waterbuck Population

At Lake Nakuru National Park, Kenya, the density of waterbuck reaches up to 250 animals per sq mi (100 per sq km) in some areas. The park average of 75 animals per sq mi (30 per sq km) is so much higher than the more typical 2.5–5 animals per sq mi (1–2 per sq km) that there are grounds for suspecting the Nakuru waterbuck differ in social behavior from other populations.

Male waterbuck usually occupy territories larger than 250 acres (100ha), but at Lake Nakuru the severity of competition probably explains why territories are smaller than elsewhere—25–100 acres (10–40ha)—and average duration of territory ownership is shorter (about 1.5 years). At any one moment, only a very small proportion (about 7 percent) of the adult males hold a territory, and only 20 percent of the males surviving to prime age ever become owners of a territory.

Fifty-three percent of the territories at Lake Nakuru contain one or more "satellite males," adult males subordinate to the territory holder and tolerated by him. Satellite males have access to the whole territory, and some have access to two adjacent territories. While tolerating his satellite(s) in the territory, a territory owner will threaten and chase out of the territory other adult males. Adult males which are neither territory holders nor satellites (over 80 percent of the adult male population) unite with young and juvenile males into bachelor herds, which rarely enter the territories of the dominant males.

Adult and young males attempting to enter a territory are often confronted and repelled by the satellite male instead of by the territory owner. Satellites apparently share in the defense of the territory and a territory holder having a satellite might thus save energy and decrease the risk of being wounded in a fight.

When a receptive female is in the territory, usually only the territory holder copulates with her. Occasionally, however, a satellite male manages to copulate with a receptive female while the territory holder is not close by. Waterbuck territories are situated along rivers and lakeshores where the grass is noticeably greener, so by being inside a territory satellite males gain access to better resources than bachelor males.

The biggest advantage for the satellite male is probably his high chance of being a territory owner himself. In 5 of 12 observed cases of change of territory ownership, a satellite became the new territory owner either on the territory it had already occupied as a satellite or on a territory adjacent to it. It can be calculated that the average probability of gaining possession of a territory is about 12 times greater for a satellite male than for a bachelor.   PW

favorite item of the diet. The primitive territorial system of the Mountain reedbuck is also found in these species. At low density, the territory of a male is shared by one resident female but home ranges were found to overlap in a high-density Zululand population, with 33 reedbuck per sq mi (16.6 per sq km). Even here, however, groups of more than three reedbuck were rare, showing that the socialization process is still strongly inhibited. Reedbuck do concentrate in much larger numbers on good pastures or when caught in open country after a fire has destroyed their normal cover of tall riverside grasses and reeds. These associations are unstable and social bonds are weak. Exceptionally high local densities of up to 285 Bohor reedbuck per sq mi (110 per sq km) occur at the end of the dry season along the upper tributaries of the Nile. In this area, groups of up to four females with attendant young are centered on the tiny territories of individual males.

Living near to rivers and lakes, but inhabiting the adjacent savanna and woodland, are the large, shaggy and slightly ungainly waterbuck. The French name for these animals is *Cobe onctueux* or "greasy kob," a reference to the oily musky secretion that covers the hairs of the coat and which is

detectable at about 1,600ft (500m) in light airs. In captivity, waterbuck require a diet with much more protein than that of other bovids, for which they require a high water intake. In the wild, this constraint accounts for their localization close to permanent water. Their diet is normally made up from short and medium grasses, reeds and rushes, but it is supplemented by browsing and wading for aquatic vegetation in the dry season. Some riverside vegetation is usually present in the waterbuck's domain. By accepting a fairly mixed diet from this rich habitat, individuals can be fairly sedentary. Female home range size varies from 0.12sq mi (0.3sq km) in a Kenyan population of density 95–140 does per sq mi, to 2.3sq mi (6sq km) in a Ugandan population of density 10 does per sq mi. The home range is typically shared by 5–8 does, although these animals do not move about as a close-knit

group; it overlaps with several of the smaller male territories, which average 32–593 acres (13–240ha) in size, depending on population density.

Waterbuck are moderately long-lived and males may hold territories from the age of 6–10 years. In East Africa, young adults of 5–6 years employ a variety of strategies to gain their own breeding territory within the permanently established network (see box).

Inhabiting the gently rolling hills and low-lying flats close to permanent water, the kob grazes on shorter savanna grass than the waterbuck. Frequently occurring at high density, the female kob moves in bands of 30–50 animals and for much of the year remains fairly sedentary. In the rainy season, herds of over 1,000 animals congregate locally, keeping the grasses short and in good growing condition for continuous cropping. In dry months, patches of green

▲ **Gestures of grazing antelopes.** (1) Southern reedbuck (*Redunca arundinum*) in the "proud posture." (2) Defassa waterbuck (*Kobus ellipsiprymnus defassa*) showing the dominance display. (3) Gray rhebok (*Pelea capreolus*) in the alert posture. (4) Uganda kob (*Kobus kob thomasi*) in the head-high approach to a female during the mating season. (5) Roan antelope (*Hippotragus equinus*) in the submissive posture. (6) Sable antelope (*Hippotragus niger*) presenting horns, a male dominance display. (7) Addax (*Addax nasomaculatus*) performing the flehmen test after sampling a female's urine. (8) Gemsbok (*Oryx gazella*) showing the ritual foreleg kick during courtship. (9) Coke's hartebeest (*Alcephalus busephalus cokii*) showing the submissive posture of a yearling. (10) A territorial male impala (*Aepyceros melampus*) roaring during the rutting season. (11) A male bontebok (*Damaliscus dorcas dorcas*) initiating butting by dropping to his knees. (12) Topi (*Damaliscus lunatus*) in the head-up approach to a female. (13) Brindled gnu (*Connochaetes taurinus*) in the ears-down courtship approach.

grass may draw kob from long distances, resulting again in the formation of large assemblages.

Adapted to the floodplains and seasonally inundated swamps, lechwe are the most specialized of the reduncine tribe. Their diet consists principally of grasses, with a distinct preference for leaf over stem. On the Kafue flats of Zambia the herds graze along the floodline in the dry season, typically in a depth of 2–8in (5–20cm) water, but a few animals venture so deep in search of food that water covers their backs. Aggregations of several hundreds (formerly thousands) occur, and mass movements have been observed in response to heavy rains and rising flood waters. In structure, lechwe populations resemble those of the related puku and kob. Close associations between individual females have not been observed and even the calf's bond with its mother is loose. Like the Uganda kob, males of Kafue lechwe may be found in conventional territorial breeding grounds (TGs). TGs of lechwe are only temporarily manned for a few weeks during the rut. Usually they hold 50–100 males within a circular area of approximately 1,640ft (500m) diameter. Large mixed herds of lechwe have been observed adjacent to these grounds.

The close adaptation of lechwe and the related Nile lechwe to their semi-aquatic habitat makes them particularly vulnerable to organized hunting. Traditional lechwe drives or *chilas* which took place along the Kafue River in the 1950s accounted for up to 3,000 animals per *chila*, while uncontrolled hunting of the Black lechwe in northern Zambia has reduced the population to a level barely one-tenth the estimated carrying capacity. Today, dams and drainage schemes are potentially just as devastating as overhunting. The Kafue Gorge hydroelectric scheme, which was completed in 1978, involved the building of dams at either end of the flats most favored by the lechwe. The total population of 94,000 lechwe subsequently dropped to about one half.

The fertile grasslands and woodlands of the moist southern and northern savannas are the residence of large herds of gnus, hartebeests and impala (tribe Alcelaphini). This tribe illustrates well the relationship between a population's ecology (distribution of available food, water, cover and other components of the habitat) and its social structure and mating system. Although its long face and sloping back mark the topi unmistakably as a true alcelaphine, it has comparatively generalized features. Widely distributed across the moist grasslands of Africa, it specializes on the green grass of valley bottoms and intermediate vegetation zones; sedentary populations occur within woodland, particularly where strips of woodland and grassland intersect. Depending on the population density, single bulls defend territories of 60–1,000 acres (25–400ha) which contain small groups of 10–20 females. The resident females are hierarchical and threaten, chase and even fight intruding females. Records of known individuals in the same territory for over three and a half years imply a remarkably stable group structure. In many parts of Africa, the seasonal availability of green grass is affected over large areas by annual droughts or floods. In those parts, the movement of topi between pastures is much more extensive, but follows, nonetheless, a predictable cyclic pattern. In the Akagera National Park of Rwanda, herds of up to 2,000 topi sweep over pastures on the broad valley bottoms. As the mating season draws near, males aggregate, up to 100 at a time, on traditional grounds similar to those of the Uganda kob. Each male defends a small territory or stamping ground which may be reduced to 80ft (25m) diameter in the densest clusters. The topi herds of the Ruwenzori National Park, Uganda, which number 3,000–4,000 animals, exploit a well-

defined area of 30sq mi (80sq km) of open grasslands. The animals apparently behave as a single population with no internal structure, and all individuals move freely over the entire plain. Since the herds have no regular directional movement, the bulls are not found on TGs. Instead, they stay with the main aggregations, and constantly engage in efforts to herd together "wards" of females and chase out other males. A typical ward is 265–330ft (80–100m) in diameter and attached to a specific piece of ground. In the middle of an aggregation, a ward contains 30–80 females, but as the herd moves on the wards slowly empty.

Though less selective feeders of medium and long savanna grassland, hartebeest are particularly fond of the edges of woods, scrub and grassland. Although basically sedentary, the female hartebeest in Nairobi National Park, Kenya, are locally quite active, moving in small groups within individual home ranges of 1.4–2.1sq mi (3.7–5.5sq km). At an overall density of 57 hartebeest per sq mi, these ranges are too large for male defense. Bulls therefore defend small territories of average size 0.21sq mi (0.31sq km), which are not particularly associated with any one female group. In

▲ **Water-chase.** Lechwe running through a floodplain in Botswana. Lechwe are the most aquatic of the grazing antelopes.

▶ **With horns interlocking,** two male Uganda kobs dispute dominance. No discrete social units or stable associations have been recorded in either male or female kobs; however, it appears that herds are oriented about traditional breeding grounds (abbreviated "TGs"). In western Uganda. TGs contain 10–20 males on closely packed central territories of 50–115ft (15–35m) diameter, surrounded by a similar number of slightly larger and more widely spaced peripheral territories. Females visit TGs for mating all the year round, usually moving directly towards the central territories, triggering off a chain reaction of clashes and ritualized displays among the males. The whole TG is usually 650–1,310ft (200–400m) in diameter and situated on smooth, slightly raised ground which is well trampled and grazed short. Some TGs have been observed on the same spot for over 15 years and traced back a further 30 years through local inhabitants.

▷ **Migratory wildebeest crossing a river.** OVERLEAF There are still vast herds of the Brindled gnu or Blue wildebeest from northern South Africa to Kenya. Such a herd on the move is a dramatic sight. Wildebeest often perish in large numbers at river crossings.

fact, the average female home range includes over 20–30 male territories.

The open grasslands and woodlands of the southern savannas are also exploited by the bizarre-looking wildebeest or gnu, which are particularly common in areas where pruning, by fire and other herbivores, has maintained a short sward of grass. As with the topi, wildebeest populations may be either sedentary or nomadic, depending on the local distribution of rainfall and green grass. Again corresponding to the topi, discrete small herds of female wildebeest occur in sedentary populations, together with a permanent territorial network of bulls and segregated bachelor groups. However, wildebeest herds are not known to be closed to outsiders, nor do the bulls associate exclusively with a single female group.

No social structure has been detected among the vast nomadic assemblages of the Brindled gnu. Both sexes are present, and during the rut males establish temporary territories whenever the aggregations come to rest. These small territories are firmly attached to a fixed piece of ground but are seldom held for more than 10 hours.

The diet of cattle is broadly similar to that of the Alcelaphini, which consequently are viewed by livestock owners as competitors for dry-season forage. Over the past few decades, the total range of the alcelaphines has severely contracted with the increase in numbers of livestock, and several populations have been nearly exterminated. Particularly vulnerable is the hirola, a species confined to a small region of dry savanna on either side of the border between Kenya and Somalia. Its total Kenyan population declined from about 10,000 in 1973 to 2,385 in 1978. Concomitantly, the number of cattle sharing the same range increased from 200,000 head to 454,414 head. The vulnerability of the alcelaphines to loss of habitat is brought out by the recent history of the bontebok. This race also has a restricted coastal range, in this case along the southwestern Cape. Indiscriminate hunting and the enclosure of the best land for farms by the early settlers had already seriously reduced the numbers by 1830. In an attempt to avoid the inevitable path to extinction, the first Bontebok National Park was established in 1931 and stocked with the pathetic total of 17 animals. This move and the creation of a second Bontebok National Park in 1961 removed the subspecies from danger; at the end of 1969 the total number in the whole Cape Province was about 800.

The dominant antelope of the less fertile woodlands of central and southern Africa is the impala. Golden tan herds of 100 or more animals moving through the park-like woods provide an unforgettable sight. Impala are selective but opportunistic feeders, accepting a broad range of dietary items including grasses, browse leaves, flowers, fruits and seeds. In the Sengwa Research Area of northwestern Zimbabwe, their diet changed from 94 percent grass in the wet season to 69 percent herbs and woody browse in the dry season. Throughout their range, impala prefer zones between different vegetation types and are particularly abundant along evergreen riverside strips in the dry season. Here, a wide variety of plant species is available and the impala can meet their annual forage requirements in home ranges of 0.2–1.7sq mi (0.5–4.5sq km). Female group size varies according to the season, averaging between 7 and 33 impala in the Sengwa area, but despite this wide variation, the population has a distinctive clan structure. Male society is much looser and essentially independent from that of the females, except with regard to the social organization in mating.

The reproductive cycle of impala is closely linked to the annual pattern of rainfall. In the equatorial region, births occur in all months, but are concentrated around two peaks associated with two rainy seasons. In contrast, a single well-defined peak of births, lasting for two to three weeks, is observed in southern Africa, which has a single wet season. The difference in timing of breeding has far-reaching effects. In both regions, impala males are territorial and, in the manner of hartebeest, they defend a smaller area than that used by an individual female. In equatorial regions, however, territory size is up to five times larger and male tenure of territory is ten times longer, leading to a prolonged displacement of bachelor groups from the best pastures. Dominance interactions are less likely to lead to fights in the equatorial zone, and territories change ownership less frequently. The short intense rut of the southern populations is heralded by displays of contagious chasing and roaring among males, not dissimilar to (although less ritualized than) oryx tournaments. At its peak, up to 180 roars per hour have been logged during the impala rut. Males mobilize all their fat reserves for this impressive breeding effort, but although fights are not uncommon, serious injury is surprisingly rare. Six to seven months after the rutting peak, single impala lambs are born. It is now the mother's turn to mobilize her fat reserves to meet the demands of lactation.

Half the newborn impala are taken by predators in the first few weeks of life, and this high toll is a clue to the close synchronization of breeding of the mothers. Preliminary evidence suggests that the timing of conceptions in impala and wildebeest can be synchronized with the phase of the moon. By ensuring that her offspring is born at the same time as those of others, a mother takes advantage of the temporary excess which satiates the local predator community. Other social factors probably contribute to the advantage of synchronized breeding.

The dry regions of Africa are the province of the Hippotragini, a tribe of horse-like antelopes. The driest country of all is inhabited by the addax, a large white antelope with magnificent spiraling horns and an elegant chestnut wig. Surviving in waterless areas of the Sahara, particularly in dune regions, the addax is well adapted to heat, coarse foods and the absence of water. They are reputed to have a remarkable ability to sense patches of desert vegetation at long distance, and apparently obtain sufficient water from their diet of grasses. Greatly enlarged hooves and a long stride improve travel over the sandy and stony desert soils.

The sparsely vegetated subdesert and Sahelian steppes of North Africa are inhabited by the Scimiter oryx, another large white antelope of the same size as the addax, but with scimitar-shaped horns and russet markings. Similar habitats in the Arabian and Sinai peninsulas were once populated by the smaller Arabian oryx (see pp132–133). The distinctively marked Beisa oryx inhabits the short-grass savannas and dry open bushland in seasonally arid areas of the Horn of Africa, and the closely related gemsbok is distributed over the dry plains and subdesert of the kalahari. These semidesert species supplement a basic diet of grass with browse: acacia pods, wild melons, cucumbers, tubers and the bulbs of succulents.

The arid environments of the oryx and addax have given rise to a remarkably tight social structure. The typical herd numbers less than 20 (60 in the gemsbok) and may contain several adult males in addition to females and young. The herd is closed to outsiders, horns providing females with the means to exclude competitors from scarce resources. Temporary aggregations of several hundred animals occur in areas where rainstorms have brought on the vegetation, but it is probable that groups typically inhabit conservative clan home ranges. Within the group, both bulls and cows are hierarchically organized; gener-

ally bulls rank over cows, but sometimes subadult bulls are dominated by high ranking cows. While the alpha bull of an oryx herd frequently reinforces his position over the other males through dominance encounters, he does not fight them and (in a complete reversal of the typical pattern among hoofed mammals) may actively prevent subordinate bulls from leaving.

The demands of a hierarchical society, coupled with the risk of conflict with long sabre-like horns, have given rise to the unique and highly ritualized oryx tournaments, which may involve many members of the herd running around in circles with sudden spurts of galloping and ritualized pacing interspersed with brief horn clashes.

Formerly protected by their inhospitable environment, addax and Scimitar oryx have become easy prey to mounted and motorized hunters with modern firearms. Sadly, the Scimitar oryx is now extinct north of the Sahara. Each year, deep boreholes for watering cattle on the edge of the Sahel are dug further north and each year the southern population of Scimitar oryx gets rarer, as cattle consume their traditional pastures. The Red Data Book states that the addax and Scimitar oryx are in real danger of total extinction in the very near future. The Arabian oryx was totally exterminated in the wild in the 1970s, but saved from extinction by the intervention of the Fauna and Flora Preservation Society and the Oman government (see pp132–133). Addax and oryx are easily tamed and were domesticated by the ancient Egyptians. A domestic herd of oryx is now being run successfully on a cattle ranch in Kenya.

Sharing the same tribal name as the oryx,

▲ **Male and female Sable antelopes,** showing a pronounced sexual difference in coat color. The male is administering the foreleg kick, a courtship gesture.

▶ **Abundance in a National Park.** A mixed group of gemsbok and springbuck and a vast flock of Red-billed quelea in Etosha National Park, Namibia.

▼ **The impala clan.** The basic social unit of impala is the herd or clan of up to 100 females and young. The clan is frequently broken up into several groups which are unstable in composition. Males generally leave the herd before they reach breeding age. During the mating season, territorial males, as here, attempt to control groups of females and young that enter their territory.

the thickly-maned Roan and Sable antelopes inhabit the moister grasslands and open woodlands. Their tribal affiliation shows up in a preference for dry hillsides over fertile river valleys. Roan antelope in Kruger National Park, South Africa, move in herd ranges averaging 31sq mi (80sq km) with minimal range overlap. Their social organization is similar to that of oryx, with closed groups of 4–18 females, but with only one adult male per group. Aggregations of up to 150 occur in the dry season, when births are rare. Sable antelope particularly enjoy woodland and grassland near to water points and are the least conditioned by their dry-country origins. The herd home range may cover an area anywhere from 3.8–124sq mi (10–320sq km) in size on an annual basis, but is confined to as little as 0.9sq mi (2.5sq km) during the breeding season. The movements of Sable antelope at this time are sufficiently localized and predictable for dominant males to relinquish their positions as masters of a mobile harem in preference for that of owner of a territory. In Zimbabwe, territories of 62–100 acres (25–40ha) have been held continuously for two or more years. During the two-month rutting peak, breeding groups of 10–20 are rounded up by territorial males who turn escaping females back to their borders with loud snorts and vicious horn sweeps.

The last member of the tribe of horse-like antelope vanished in about 1800. Standing in the path of the pioneer columns, the bluebuck of the Cape Province (a close relative of the Roan antelope) fell prey to hunters and sportsmen before the concept of conservation had been born. Mounted specimens are in museums in Leiden, Paris, Stockholm, Vienna and Uppsala University.

MGM

THE 24 SPECIES OF GRAZING ANTELOPE

Tribe Reduncini

Africa. Antelopes of wetlands, tall or tussock grassland. Horns on male only; long hair on sides and neck except in kob.

Genus *Redunca*

Light and graceful animals characterized by whistling and high bouncing jumps. Females of similar size to males. Glandular spot occurs beneath ears. 3 species.

Southern reedbuck
Redunca arundinum

Africa north to Tanzania and west to Angola. Southern savannas. Largest reedbuck. HBL 53–66in; TL 10–11in; SH 33–38in; HL 12–18in; wt 110–176lb. Coat: light buff to sandy brown; pale zone at base of horns; black and white markings on forelegs. Subspecies: 2.

Mountain reedbuck
Redunca fulvorufula

Cameroun, Ethiopia and E Africa, S Africa. Montane grasslands up to 16,400ft. Smallest reedbuck. HBL 43–53in; TL 8in; SH 25–30in; HL 5–15in; wt 66lb. Coat: soft woolly and gray. Prominent eyes and sockets. Subspecies: 3, each confined to its own highland area, including **Chanler's mountain reedbuck** (*R.f. chanleri*), Ethiopia and E Africa.

Bohor reedbuck
Redunca redunca

Senegal east to Sudan and south to Tanzania. Northern savannas. HBL 39–51in; TL 7–8in; SH male 29–35, female 27–30in; HL 8–16in; wt 99lb. Coat: light buff with strongly marked forelegs. Horns forward hooked. Subspecies: 7, including the Sudan form (*R.r. cottoni*) with long splayed horns.

Genus *Pelea*

A single species.

Gray rhebok†
Pelea capreolus
Gray rhebok, Vaal ribbok or rhebuck.

S Africa. High plateaus. SH 30in; HL 8–11in; wt 51lb. Gracefully built with long slender neck. Coat: brownish gray, soft and woolly. Tail short and bushy. No gland patch beneath the ears. Horns short, straight and almost vertical.

† This species is now recognized as not being a reduncine. Its exact affinities are still under debate.

Genus *Kobus*

Medium- to large-sized animals with a relatively heavy gait. Males have ridged horns and females are smaller than males. 5 species.

Waterbuck
Kobus ellipsiprymnus

S Africa north to Ethiopia and S Sudan, west to Senegal. Savanna and woodland near to permanent water. HBL 70–92in; TL 13–16in; SH 19in; HL 21–39in; wt 374–550lb. Big and shaggy with heavy gait. Coat: dark gray or reddish. Two taxonomic groups. The **Common waterbuck**, *ellipsiprymnus* group, with a white ellipse on the rump, is restricted to the east of the rift valley for most of its range extending from 6°N to 29.5°S. Subspecies: 4. The **Defassa waterbuck**, *defassa* group, with a white blaze on the rump is predominantly reddish in color. North and west of the rift valley. Subspecies: 9.

Kob
Kobus kob

Gambia east to Sudan and Ethiopia, Uganda. Low-lying flats and gently rolling hills close to permanent water. HBL 63–71in; TL 4–6in; SH male 35–39in, female 32–36in; HL 16–27in; wt 170lb. Male robust and thick-necked with lyre-shaped horns; females more slender and graceful. Coat: reddish with white underside and throat chevron; legs with black markings. Subspecies: 10, including **Buffon's kob** (*K.k. kob*), Senegal to NW Nigeria; **White-eared kob** (*K.k. leucotis*), of S Sudan, whose males are predominantly black.

Lechwe [V] [*]
Kobus leche

Botswana, Zambia, SE Zaire. Floodplains and seasonally inundated grasslands. SH 39in; HL 20–36in; wt male 275lb, female 155lb. Coat: long and rough, bright chestnut to black with white underparts. Hooves long and pointed. Horns long, thin and lyre-shaped. Subspecies: 3. **Red lechwe** (*K.l. leche*), Botswana and Zambia. **Black lechwe** (*K.l. smithemani*), NE Zambia and SE Zaire. **Kafue Flats lechwe** (*K.l. robertsi*), Kafue River, Zambia.

Nile lechwe
Kobus megaceros
Nile or Mrs Gray's lechwe.

Sudan, W Ethiopia. Swamps in the vicinity of the White Nile, Sobat, Baro and Gilo Rivers. SH 37in; HL 24–34in; wt 190lb. Coat: long and rough, blackish chocolate with white underparts and white patch on shoulders; female uniformly reddish fawn. Hooves long, pointed and splayed. Horns long and spread out.

Puku
Kobus vardoni

S Zaire, Botswana, Angola, Zambia, Malawi, Tanzania. Margins of lakes, swamps and rivers, and on floodplains. HBL 50–60in; TL 11–12in; SH 30–33in; HL 16–21in; wt male 170lb, female 145lb. Coat: fairly long and shaggy, bright golden yellow. Horns short, thick and less lyre-shaped. Subspecies: 2. **puku** (*K.v. vardoni*) and **Senga kob** (*K.v. senganus*).

Tribe Alcelaphini

Africa. Grazers of open woodland, moist grassland and the zone between these two habitats. Characteristically with high population density. Except for impala, tribe characteristics include horns on both sexes, a long face, elevated shoulder and sloping hindquarters, preorbital glands and glands on forefeet only. The female is slightly smaller than the male. Impala have commonly been classed with the gazelles (Antelopini) although affinity with the kobs and reedbuck (Reduncini) has also been suggested. Evidence now suggests that impala are an early offshoot of the Alcelaphini, having diverged at the start of that tribe's history as an independent unit.

Genus *Beatragus*

A single species.

Hirola [R]
Beatragus hunteri
Hirola or Hunter's hartebeest.

E Kenya and S Somalia. Grassy plains between dry acacia bush and coastal forest. HBL 47–49in; TL 12–18in; SH 39–49in; HL 22–28in; wt 176lb. Coat: uniform sandy to reddish tawny, with white spectacles. Horns long and lyre-shaped on short bony base (pedicel). Female smaller with lighter horns.

Genus *Damaliscus*

Characterized by a long head but frontal region not drawn upward into a bony pedicel. Slope of body less exaggerated than that of hartebeest. Horns thick and ridged.

Topi [*]
Damaliscus lunatus
Topi, tsessebe, sassaby, tiangs damalisc, korrigum or Bastard hartebeest.

Senegal to W Sudan, E Africa through to S Africa. Green grassland of open savanna and swampy floodplains. HBL 67in; TL 17in; SH 49in; HL 14–24in; wt male up to 375lb, female 286lb. Coat: bold pattern of black patches (notably on face) on glossy mahogany red; fawn on lower part of legs. Subspecies: 7, including **tsessebe** (*D.l. lunatus*), from Zambia southwards; *D.l. jumela* and **topi** (*D.l. topi*) in NE Zaire and E Africa; **tiang** (*D.l. tiang*) in NW Kenya, W Ethiopia, and S Sudan; and **korrigum** (*D.l. korrigum*) from Senegal to W Sudan.

Bontebok
Damaliscus dorcas
Bontebok or blesbok.

S Africa. Open grassland. SH 33–39in; HL 14–20in; wt 130–220lb. Coat: Rich purplish chestnut brown, darker on the neck and hindquarters; white face patch, rump, belly and lower legs. Horns rather small in simple form. Subspecies: 2. **bontebok** (*D.d. dorcas*); **blesbok** (*D.d. phillipsi*).

Genus *Alcelaphus*

Characterized by a long narrow hammer-shaped head, heavy hooked and ridged horns and pronounced slope of back. 2 species.

Hartebeest
Alcelaphus buselaphus
Hartebeest or kongoni.

Coarse grassland and open woodland. Senegal to Somalia, E Africa to S Africa. HBL 77–79in; TL 12in; SH 44–51; HL 18–27in; wt male 313–403lb, female 313–370lb. Coat: uniform sandy fawn to bright reddish, lighter on hindquarters, sometimes with black markings on legs. Frontal region of head drawn up into a bony pedicel. Horn shape diagnostic of races. Subspecies: 12, including **bubal** or **Northern hartebeest** (*A.b. buselaphus*) ranged N of Sahara, extinct; **Western hartebeest** (*A.b. major*), Senegal and

Guinea; **Lelwel hartebeest**
(*A.b. lelwel*), S Sudan, Ethiopia, N
Uganda and Kenya; **Tora hartebeest**
(*A.b. tora*) [E], Sudan, Ethiopia;
Swayne's hartebeest (*A.b. swaynei*)
[E], Ethiopia, Somalia; **Jackson's
hartebeest** (*A.b. jacksoni*), E Africa,
Rwanda; **Coke's hartebeest** (*A.b.
cokii*), Kenya, Tanzania; **Cape** or **Red
hartebeest** (*A.b. caama*), S Africa,
Namibia, Botswana, W Zimbabwe.

Lichtenstein's hartebeest
Alcelaphus lichtensteini

Tanzania, SE Zaire, Angola, Zambia,
Mozambique, Zimbabwe.
Open woodland. HBL 75in;
TL 18in; SH 49in; HL 18–24in;
wt male 352–452lb, female 365lb.
Coat: bright reddish with fawn flanks
and white hindquarters; dark stripe
down front legs. Frontal region of
skull does not form pedicel.

Genus *Connochaetes*

Characterized by massive head and
shoulders with mane on neck and
shoulders, beard under throat and
long tail reaching nearly to the
ground. Both sexes horned. 2 species.

White-tailed gnu
Connochaetes gnou
White-tailed gnu or Black wildebeest.

S Africa. Open grass veld. SH 45in;
HL 21–29in; wt 330–400lb. Coat:
dark brown to black; tail white. Face
covered by brush of stiff upward-
pointing hairs, tuft of hair between
front legs. Horns descending forwards
then pointing upwards. Name derived
from the hottentot '*t*' *gnu* which
describes the typical loud bellowing
snort.

Brindled gnu
Connochaetes taurinus
Brindled gnu or Blue wildebeest.

Northern S Africa to Kenya just south
of the equator. Moist grassland and
open woodland. HBL 76–82in;
TL 18–22in; SH 50–55in;
HL 16–29in; wt male 510lb, female
352lb. Bovine appearance. Coat:
slaty to dark gray; tail black. Horns
curving downwards laterally and
then pointing upwards and inwards.
Subspecies: 5, including **Blue
wildebeest** (*C.t. taurinus*), Zambia and
to the south and west; **Cookson's
wildebeest** (*C.t. cooksoni*), Luangwa
valley, Zambia; **White-bearded
wildebeest** (*C.t. mearnsi*), Tanzania,
Kenya.

Genus *Aepyceros*
A single species.

Impala
Aepyceros melampus

S Africa to Kenya, Namibia to
Mozambique. Open deciduous
woodland, especially near water.
HBL male 56in, female 50in;
TL 12in; SH male 36in, female
34in; HL 19in; wt male 176lb,
female 95lb. Straight backed, light
limbed, and graceful; the
quintessential antelope. Coat: light
mahogany with fawn flanks and
white undersurface of the belly; black
vertical stripes on tail and thighs, and
black glandular tuft on the fetlocks.
Horns on male only, lyre-shaped.
Subspecies: 6, including **Black-faced
impala** (*A.m. petersi*) [E], SW Angola,
NW Namibia.

Tribe Hippotragini
Horse-like antelope of dry country
and savanna. Africa, Arabia. Females
only marginally smaller than males
and both sexes with well-developed
horns. 7 species (1 extinct).

Genus *Hippotragus*
Characterized by heavily ringed horns
curving backwards. Well-developed
mane of stiff hairs. 3 species.

Roan antelope [*]
Hippotragus equinus
Roan or Horse antelope.

Gambia to the Somali arid zone, C
Africa to S Africa, but rarely east of
the Rift Valley. Thinly treed
grasslands. HBL 75–95in; TL 14–19in;
SH 50–57in; HL 22–39in;
wt male 620lb, female 575lb.
Coat: sandy fawn to dark
reddish with white underparts;
contrasted black and white face
markings. Ears long with tuft of hair
on tip. Subspecies: 6.

Sable antelope [*]
Hippotragus niger

C Africa from Kenya to S Africa,
Angola to Mozambique. Woodland
and woodland-grassland edges.
HBL 77–83in; TL 15–18in;
SH 46–55in; HL 19–65in; wt male
575lb, female 485lb. Coat:
adult females and young bulls rich
russet with pale underparts,
darkening to black; mature bulls
black with white bellies; calves
uniformly dun. Forehead narrow and
horns scythe-like. Subspecies: 4,
including **Giant sable antelope** (*H.n.
variani*), Angola.

Bluebuck [EX]
Hippotragus leucophaeus
Bluebuck or blaauwbok.

Formerly SW Cape Zone; extinct
about 1800. Woodland glades.
SH 41in; HL 13–24in. Coat: Bluish
gray with white underparts and
insides of limbs (the velvety-bluish
sheen of the coat was much admired);
face markings indistinct or absent.
Hairs on neck mane directed forward,
tail short and tufted.

Genus *Oryx*
Characterized by short mane, hump
over shoulder and large hooves.

Scimitar oryx [E]
Oryx dammah
Scimitar or White oryx.

Formerly over most of N Africa,
presently extinct north of the Sahara;
confined to a narrow strip between
Mauritania and the Red Sea. Sahel
and semidesert. SH 47in; HL 40–50in;
wt 450lb. Coat: pale with
neck and chest ruddy brown and
brownish markings on the face.
Scimitar-shaped horns.

Gemsbok
Oryx gazella
Gemsbok, oryx or Beisa oryx.

Discontinuous distribution. Namibia,
Angola, Botswana, Zimbabwe,
S Africa, and Tanzania north to the
Ethiopian coast. Seasonally arid
areas. HBL 60–67in; TL 18in;
SH 47in; HL 25–43in; wt male
440lb, female 360lb. Coat: fawn
with white underparts, black and
white markings on the head; black
line down throat and across flanks;
black tail. Horns straight and long.
Subspecies: 5, including **gemsbok**
(*O.g. gazella*) in SW Africa; **Beisa oryx**
(*O.g. beisa*) in the Horn of Africa; and
Fringe-eared oryx (*O.g. callotis*) in
Kenya and Tanzania.

Arabian oryx [E]
Oryx leucoryx
Arabian or White oryx.

Formerly Arabian peninsula, Sinai
peninsula. Presently reintroduced
into Oman. Stony semidesert. About
two-thirds of the size of *gemsbok*.
SH 32–40in; HL 15–27in; wt
143–165lb. Coat: white with black
markings on the face; legs dark
chocolate brown to black; tawny
colored line across flanks. Horns
nearly straight.

Genus *Addax*
A single species.

Addax [V]
Addax nasomaculatus

Formerly through entire Sahara.
Presently remnant populations in
Mauritania, Mali, Niger, Chad,
S Algeria, W Sudan. Sandy and stony
desert far from water. SH 39–43in;
HL 30–43in; wt 178–270lb. Coat:
grayish white with white rump, belly
and limbs; chestnut wig on forehead.
Tail long with black tip. Horns long
and spirally twisted.

MGM

The Arabian Oryx

A specialist for extremes

The desert environment of central Arabia is one of cruel extremes: summer shade temperatures peak at 118–122°F (48–50°C), while winter minima drop to 43–45°F (6–7°C) with strong cold winds. On every day of the year, the temperature varies by 36°F (20°C). Over large areas, rain may not fall for years and the vegetation lies dormant or in seed; natural surface water is short-lasting after rain. Sand storms can reduce visibility to a few yards for days on end. Resources of food, water, shade and shelter are sparse and scattered, making exacting demands on the animals living here. Of all the oryx species, gemsbok in the Kalahari, beisa in the East African Somali desert, and Scimitar-horned oryx in the Sahara, none equals the Arabian oryx as a desert specialist.

Sparse food has selected for a small size of 143–165lb (65–75kg), only a quarter of the weight of the gemsbok in the Kalahari which, although a barren landscape, is well vegetated in comparison with central Arabia. Its color patterning has been simplified to produce a predominantly white coat which largely reflects incoming solar radiation. On winter mornings, the animals' coat has a suede look as they erect their hairs to absorb the sun's warmth, which is retained at night by the thicker winter coat. The blacker leg markings in winter also increase heat absorption.

The hooves are splayed and shovel-like, with their large surface areas fully in contact with the ground as an adaptation to sandy surfaces. The oryx is not a great runner, but it can walk for hours on end to reach a favorable habitat, and treks of 15–19mi (25–30km) in one night are not unusual. The oryx often combines trekking with the unavoidable chore of rumination,

but if the herd encounters a patch of good grazing, they switch to feeding at once.

As the desert allows only low overall densities of oryx, the problems of finding a mate and reducing the risk of predation are solved by having only one herd type, containing all ages and approximately equal numbers of each sex. Such herds of 10–20 animals probably stay together for a considerable time.

Group stability is achieved through a linear dominance hierarchy in which there is a dominant bull to whom all other animals will defer; the second senior animal will be another bull or adult female who takes precedence over all except the dominant bull, and so on. Only the dominant male mates, so during his dominance, all calves are half-siblings. Newcomers meet with resistance in the form of threats, lunges and chases from established members of their own sex, and between males there can be severe fighting. If a newcomer persists in moving with the herd as a satellite, it is usually tolerated within 2–3 weeks, and is allowed to integrate progressively. Because both males and females feed in close proximity, every animal has to defend its food resources, and thus both sexes bear similar horns. The mixing of familiar individuals of each sex means that valuable energy is not used on the development of extravagant secondary sexual characteristics, and, thus, males and females are similar. The Arabian oryx expends less energy than other oryx in interactions between each other, and its musculature and tendon mass between the withers and skull is correspondingly reduced.

The low frequency of aggressive interactions allows animals to share scattered shade trees under which they may spend

eight of the daylight hours in the summer heat. Their small size allows them to creep under quite small acacia canopies for shade. These trees—like the oryx, natives of Africa—are also shorter in the desert. Under shade trees the oryx excavate scrapes with their fore-hooves so that they lie in cooler sand and reduce the surface area exposed to drying winds. They "fine-tune" their heat regulation carefully by seeking shade earlier on hot days and not venturing out until a cooler evening breeze blows. Through behavioral avoidance of excessive heat load, precious water is not lost by panting to cool down.

The members of a feeding herd spread out until neighbors may be 165–330ft (50–100m) apart, but constant visual checking, especially in undulating terrain, ensures that the animals keep in touch. Cohesion is helped by strong synchronization of activity within the herd. When the herd is trekking, a subsdominant male leads, up to 330ft (100m) in front. Changes of direction when feeding can be initiated by any adult female. She will start in the new direction, then stop and look over her shoulder at the others until more or, gradually, all start to follow her.

Separation from the herd seems accidental. Singly oryx search for their herd and can recognize and follow fresh tracks in the sand. Moreover, as oryx are visible to the naked eye at 2mi (3km) in sunlight, the white coat may have evolved partially as a flag to assist herd-location in a open environment where merging with the environment is less necessary. Non-human predators such as the Arabian wolf and Striped hyena have never been abundant.

Historically, the Arabian oryx ranged through Arabia, up through Jordan and into Syria and Iraq. It had always been a prized trophy and source of meat for bedu tribesmen who, hunting on foot or from camel with a primitive rifle, were unlikely to have significantly depleted the populations. But from 1945, motorized hunting and automatic weapons caused a severe contraction of its range and numbers, and it became extinct in 1972, leaving a few in private collections in Arabia and the World Herd in the USA. This herd grew out of the 1962 Operation Oryx, organized by the far-sighted Fauna and Flora Preservation Society to ensure the survival of the species in captivity. By 1982, ecological and social conditions in central Oman were deemed right for the release of a carefully developed herd, with the long-term aim of re-establishing a viable population.   MSP

▲ **Peace in the shade.** Arabian oryx cope with their inhospitable surroundings by reducing unnecessary energy expenditure. They are extremely tolerant of each other, thus avoiding wasteful fights, and happily share the precious shade afforded by trees and bushes.

◄ **The patient trek** of Arabian oryx moving to a grazing area in late afternoon. They are highly organized and methodical in their movements, with a subdominant male leading the group and the dominant male at the rear, rounding up the calves.

GAZELLES AND DWARF ANTELOPES

Subfamily: Antilopinae
Thirty species in 12 genera.
Family: Bovidae.
Order: Artiodactyla.
Distribution: Africa, Middle East, Indian subcontinent, China.

Habitat: varied, from dense forest to desert and rocky outcrops.

Size: head-and-body length from 18–21.5in (45–55cm) in the Royal antelope to 57–68in (145–172cm) in the Dama gazelle; weight from 3.3–5.5lb (1.5–2.5kg) in the Royal antelope to 88–188lb (40–85kg) in the Dama gazelle.

Dwarf antelopes
Tribe: Neotragini.
Twelve species in 6 genera.
Species include: **Günther's dikdik** (*Madoqua guentheri*), **klipspringer** (*Oreotragus oreotragus*), **oribi** (*Ourebia ourebi*), **Pygmy antelope** (*Neotragus batesi*), **Royal antelope** (*Neotragus pygmaeus*), **steenbuck** (*Raphicerus campestris*).

Gazelles
Tribe: Antilopini.
Eighteen species in 6 genera.
Species include: **blackbuck** (*Antilope cervicapra*), **dibatag** (*Ammodorcas clarkei*), **Dorcas gazelle** or **jebeer** (*Gazella dorcas*), **gerenuk** (*Litocranius walleri*), **Goitered gazelle** (*Gazella subgutturosa*), **Grant's gazelle** (*Gazella granti*), **Red-fronted gazelle** (*Gazella rufifrons*), **Speke's gazelle** (*Gazella spekei*), **springbuck** (*Antidorcas marsupialis*), **Thomson's gazelle** (*Gazella thomsoni*).

▶ **Pronking springbuck** fleeing from a predator ABOVE. Pronking involves a sudden vertical leap in the middle of fast running.

▶ **Kirk's dikdik,** one of the smaller dwarf antelopes. Dwarf antelopes are well endowed with scent-glands and the preorbital gland of this male dikdik is especially prominent.

THE gazelles and dwarf antelopes make up a tantalisingly contrasting group, containing some of the most abundant and the rarest, the most studied and the least known of the hoofed mammals. They occupy habitats from dense forest to desert and rocky outcrops, and their range spans three zoogeographic regions from the Cape to eastern China.

The dwarf antelope tribe (Neotragini) is very varied in form and habitat: from pygmy antelopes in the dense forests of central Africa to the oribi in the open grass plains adjacent to water; from the dikdiks in arid bush country to the klipspringer on steep rocky crags. The only common features of the tribe are their small size, with females 10–20 percent larger than males, and well-developed glands for scent marking, especially preorbital.

Their small size is associated with their diet. All the species except the oribi, which is a grazer, are "concentrate selectors," taking easily digested vegetation low in fiber, such as young green leaves of browse and grass, buds, fruit and fallen leaves. A smaller body size increases the metabolic requirement per pound of body-weight, so a small herbivore has to assimilate more food per pound of body-weight than a large her-

bivore. Ruminants like the dwarf antelopes have a limit to the rate of food intake which is dictated by the length of time food stays in the rumen. In dwarf antelopes this limit is so low that they have to choose vegetation of a high nutritive quality. Small size is evidently a secondary adaptation since their gestation time is more typical of the larger hoofed mammals the size of the gazelles. The reduction in the extent of forest habitat since their evolution in the Miocene (about 12 million years ago) has meant that smaller size has been increasingly favored.

Unusually among hoofed mammals, female dwarf antelopes are larger than the males. It is advantageous for the females to be as large as possible, within the constraints of their habitats, since they have the added burden of raising young. However, many dwarf antelopes are territorial, and among some territorial species inter-male competition for females favors larger males. But in many dwarf antelopes, the males form a lifetime bond with one or a few females, and this reduces inter-male conflict and therefore precludes the need for larger size.

Territoriality has been favored in the evolution of dwarf antelopes: because the food items are so varied and scattered

through the habitat, it is most efficient for an animal to have an intimate knowledge of its home range and the distribution of resources therein, and to exclude competitors. A permanent bond between the individuals of a breeding group, and the demarcation of an exclusive territory, help to maintain such a system.

In order to demarcate and maintain these territories, dwarf antelopes have well-developed scent glands, whose secretions have become more important than visual displays and encounters. Secretions from the preorbital glands in front of the eyes are daubed repeatedly on particular stems, where a sticky mass accumulates. Pedal glands on the hooves mark the ground along frequently traveled pathways. Males also mark females in this way, thus reinforcing the bond.

The number of scent glands is highest in the oribi, which has six different gland sites, including one below the ear. Oribi have large territories 0.4sq mi (1sq km) in open grassland, and their well-developed glands may be a response to the need to increase the amount of marking.

In dwarf antelopes dung and urine are deposited on particular sites, and the male has a characteristic posture and linked urination/defecation when adding to these piles. In most species, scent marking is done by both sexes. Typically, when a female defecates on these piles, the male will sniff, paw, and add his own contribution to hers immediately after. Oribi and suni adopt a ritualized posture and perform preorbital marking when neighboring territorial males meet; fights are very rare. Dikdiks "horn" the vegetation and raise their crests as a threat display.

The range of the dwarf antelopes has almost certainly been affected by disturbance to the habitat, since many of the species prefer the secondary growth that invades disturbed areas, notably from slash-and-burn cultivation. The pygmy antelopes, dikdiks, grysbucks and steenbuck all favor this habitat.

Those species that live in dense cover, such as the pygmy antelopes, have a crouched appearance, with an arched back and short neck, a body shape that is suited to rapid movement through the dense vegetation. The other species live in more open habitat, where detection of predators by sight is more important, and they have a more upright posture, with a long neck and a raised head. The exception is the klipspringer, which has an arched back, enabling it to stand with all four limbs together, to take advantage of small patches of level rock. Its physical appearance is unusual, having a thick coat of lightweight hollow-shafted hair to protect the body against the cold and physical damage from the bare rock in its exposed habitat of rocky crags. It has peg-like hooves each with a rubbery center and a hard outer ring to gain a sure purchase on the rock.

The proboscis of the dikdiks, which is most pronounced in the Günther's dikdik, is an adaptation for cooling. Venous blood is cooled by evaporation from the mucous membrane into the nasal cavity during normal breathing or under greater heat stress from nasal panting.

Dwarf antelopes generally rely on con-cealment to escape detection and thereby minimize predation. Their first response on detecting a predator is to freeze, and then, on the predator's closer approach to run away. Grysbucks tend to dash away and then suddenly drop down to lie hidden from sight. In tall grass, oribi lie motionless to escape detection, but, if the grass is short they flee, often with a stotting action like that of gazelles. The oribi and the dikdiks have a whistling alarm call.

Births occur throughout the year, but there are peaks that coincide with the vegetation flush that follows the early rains. In equatorial regions where there are two rainy seasons a year, two birth peaks occur. Young are born singly. Females become sexually mature at six months in the smaller species and ten months in the larger, and males become sexually mature at about fourteen months. The newborn lie out for the first few weeks after birth, and the mother will come to suckle it, calling with a soft bleat which the infant answers. When the young become sexually mature, they are increasingly threatened by the territorial male. The young responds with submissive displays such as lying down, and it eventu-ally leaves the territory at 9–15 months old.

In contrast to the dwarf antelopes, gaz-elles (tribe Antilopini) are all very similar in shape, with only two exceptions: the gerenuk (see pp142–143) and dibatag. They have long legs and necks and slender bodies, and most have ringed (annulated) S-shaped horns. They are generally pale fawn above and white beneath. The exception is the blackbuck of India, in which the males develop conspicuous black upperparts at about three years of age. The two-tone coloration of gazelles probably acts as coun-tershading to obscure the animal's image to a predator and minimize detection.

Gazelles live in more open habitat than the dwarf antelopes and so rely more on visual signals. Some species, such as the Thomson's gazelle, Speke's gazelle, Red-fronted gazelle and the springbuck, have conspicuous black side-bands. Three func-tions have been suggested for these bands: they act as a visual signal to keep the herds together, they communicate alarm when all the members are fleeing, and they may break up the outline of individuals in a herd. All the species have white buttocks with at least some black on the tip of the tail, and dark bands that are more or less distinct, being most conspicuous in Grant's gazelle. Another feature of the gazelles is their stotting or pronking gait, seen when they are apparently playing or alarmed: bounc-ing along stiff-legged with all four limbs landing together. The possible functions of this are that it communicates alarm, gives the animal a better view of the predator, and also confuses or even intimidates it. Pron-king is most pronounced in the springbuck, hence its name; it also erects the line of white hairs on its back at the same time.

The sense of sight and hearing are well-developed and are reflected in the large orbits and ear cavities of the skull. In contrast to the dwarf antelopes, most females possess horns, the exceptions being the Goitered gazelle, blackbuck, gerenuk and dibatag. One theory of horn evolution is that they have evolved in response to terri-toriality and increased inter-male conflict, and females have evolved horns to defend their food resources. Patchily distributed food resources can be defended, and this is particularly so for gazelles in the dry season or in winter, when the nutritive quality of the vegetation is generally poor, and patches of more nutritious grass and bushes occur.

The range of the tribe is very extensive, from the springbuck in the Cape to the blackbuck and Dorcas gazelle in India, and Przewalski's gazelle in eastern China. The species in North Africa, the Horn, and Asia inhabit arid and desert regions where their distribution is very patchy and they occur in low densities, with populations separated by geographical barriers of uninhabitable ter-rain, mountain ranges, and seas such as the Persian Gulf and the Red Sea. This has led to many variations in form. A good example is the Dorcas gazelle: in Morocco it is a pale-colored gazelle with relatively straight par-allel horns and as it extends eastwards it goes through various discrete changes, becoming a reddish color, with a smaller body size and lyre-shaped horns in India.

Gazelles are mixed feeders, though they

▶ **Species of dwarf antelopes and gazelles.**
(1) Klipspringer (*Oreotragus oreotragus*) defacating on its territorial boundary.
(2) Dibatag (*Ammodorcas clarkei*) in the alarmed posture. (3) Slender-horned gazelle (*Gazella leptoceros*), the palest gazelle. (4) Oribi (*Ourebia ourebia*) scent marking a grass stem with its ear gland. (5) Kirk's dikdik (*Madoqua kirkii*) horning vegetation. (6) Royal antelope (*Neotragus pygmaeus*), the smallest antelope. (7) Steenbuck (*Raphicerus campestris*) scent marking with the pre-orbital gland. (8) Tibetan gazelle (*Procapra picticaudata*). (9) Springbuck (*Antidorcas marsupialis*) pronking. (10) Goitered gazelle (*Gazella subgutturosa*). (11) Dama gazelle (*Gazella dama*), the largest gazelle.
(12) Blackbuck (*Antilope cervicapra*) in territorial display gesture.

THE LIFE OF THE THOMSON'S GAZELLE

▲ **The typical stance** of fighting gazelles is displayed by these Thomson's gazelles: heads as close to the ground as possible and horns interlocked.

◄ **Mating.** The males establish territories during the breeding season and mate with receptive females who enter it.

► **A newborn Thomson's gazelle,** with the cord still visible. The young lie out for the first few weeks until they can run with the herd.

▷ **Meal fit for a cheetah.** The not uncommon end of a gazelle's life: one graceful creature devoured by another.

mainly browse, taking grass, herbs and woody plants to a greater or lesser degree, depending on their availability. Generally, they eat whatever is greenest, so in the early spring or rains when the grass flushes they will turn to browse. Thomson's gazelle is almost entirely a grazer, up to 90 percent of its food being grass. The gerenuk is a browser that feeds on young shoots and leaves of trees, particularly acacias (see pp142 143), up to a height of 8.5ft (2.6m). The dibatag also feeds in this manner and shows similar adaptations of body shape.

In all of the gazelles, the males establish territories, at least during the breeding season, from which they actively exclude other mature males while females are receptive. Four types of groupings occur: single territorial males; female groups with their recent offspring, usually associated with a territorial male; bachelor groups comprising nonbreeding males without territories; and mixed groups of all sexes and ages that are common outside the breeding season. There is a certain amount of mixing of the groups. Males mark their territories with urine and dung piles and, in most species, with secretions from their preorbital glands. Subordinate males will also add to the dung piles when they move through the territories, and territorial males will tolerate familiar subordinate males within their territories as long as they remain subordinate and do not approach the females with any intent. Threat displays start at the lowest level of intensity with head raised, chin up and horns lying along the back. This display reaches its most advanced form in the Grant's gazelle. Two males stand antiparallel with each other, chins up and heads turned away, and at the same instant both whip their heads around to face each other. The next level of intensity is a head-on approach with chin tucked in and horns vertical. The next level is head lowered to the ground and horns pointing towards the opponent.

Fights, when they develop, consist of opponents interlocking horns with their heads down on the ground and pushing and twisting. At any time, an individual can submit by moving away at a greater or lesser speed, depending on the closeness of the visitor's pursuit. Sparring bouts are very common between bachelor males; these do not result in change of status but doubtless provide experience in sizing up opponents' abilities. Intense, wounding fights do occur between neighboring territorial males. Territorial males will herd females with a chin-up display, but if females decide to leave a

territory the male will not stop them or attempt to retrieve them once they have crossed the boundary.

In contrast to the dwarf antelopes, the food items of the gazelles are more abundant and continuously distributed in the habitat, so that territories can support a greater density of animals. Males can therefore afford to maintain many females in their territories. Since there are thus fewer breeding males per female, there is increased inter-male conflict for the right to establish territories and breed. Increased size of the male confers an advantage and so evolution has favored males larger than females.

In the temperate and cold zones of their geographical range, gazelles breed seasonally, so that the births coincide with the vegetation flush in spring or early rains. Females go off on their own to give birth and fawns lie out for the first few weeks like the dwarf antelopes. Once the fawns can run sufficiently well they join the group.

Many species of gazelles have been greatly reduced in numbers and range by man. The species that inhabit North Africa, the Horn of Africa and Asia have suffered most because they live in arid habitats where their densities are less than 2.6 per sq mi (1 per sq km), and where they can easily be hunted from vehicles. Domestic sheep and goats also compete for the same food plants, and access to springs has been denied them by human activity. Several of these species are endangered. In Asia, cultivation has removed many areas of essential winter range, so that the vast winter aggregations, particularly of the Goitered gazelle and the blackbuck, are now rare or absent. The springbuck, though still common, no longer occurs in migrating herds of tens of thousands, due to excessive hunting and to barring their natural movements with fences for ranching of domestic stock. BPO'R

THE 30 SPECIES OF DWARF ANTELOPES AND GAZELLES

Abbreviations: HBL = head-and-body length. TL = tail length. SH = shoulder height. HL = horn length. wt = weight.
[E] Endangered. [V] Vulnerable. [I] Indeterminate. [*] CITES listed.

Tribe Neotragini

Dwarf antelopes

Small delicate antelopes, females slightly larger than males. Horns short and straight, females hornless, except in a subspecies of klipspringer. Except for oribi, diet young leaves and buds, fruit roots and tubers, fallen leaves and green grass. Live singly or in small family groups. Males territorial. Gestation about 6 months. Territories marked with dung piles, preorbital and pedal glands, scent glands well developed. Breeding throughout the year with birth peaks in early rains. Underparts white or buffish.

Genus *Neotragus*

Horns smooth, short, inclined backwards. Tail relatively long. Back arched, neck short. 3 species.

Royal antelope
Neotragus pygmaeus

Sierra Leone, Liberia, Ivory Coast and Ghana. Dense forest. Occurs singly or in pairs, shy and secretive. Little known. Smallest horned ungulate. HBL 18–22in; TL 1½–2in; SH 8–11in; HL 1–1¼in; wt 3–6lb. Head and neck dark brown. Coat: back brown, becoming lighter and bright reddish on flanks and limbs, contrasting with white underparts; tail reddish on top and white underneath and at tip.

Pygmy antelope
Neotragus batesi

SE Nigeria, Cameroun, Gabon, Congo, W Uganda and Zaire. Dense forest. Predominantly solitary. Often move into plantations and recently disturbed land at night. Females have overlapping home ranges. HBL 20–23in; TL 1½–2in; SH 9–13in; HL ¾–1¼in; wt 6lb. Coat: shiny dark chestnut on the back, becoming lighter on the flanks; tail dark brown.

Suni
Neotragus moschatus

Patchily distributed from Zululand through Mozambique and Tanzania to Kenya. Coastal, riverine and montane forest with thick undergrowth. Occur singly or in small groups, with never more than one adult male per group. Males grate their horns on tree trunks as a territorial marker. HBL 23–24in; TL 4–5in; SH 12–16in; HL 2–5in; wt 9–13lb. Coat: dark brown, slightly freckled; transparent pink ears. Horns strongly annulated at base.

Genus *Madoqua*

Long, erectile hairs on forehead. Tail minute. Underparts white. Pairs of a male and female form lifelong territories. Larger groupings commonly occur at sites of food abundance between territories. 3 species.

Swayne's dikdik
Madoqua saltiana
Swayne's or Phillips' dikdik.

Horn of Africa. Arid evergreen scrub in foothills and outliers of Ethiopean mountains, particularly in disturbed or over-grazed areas with good thicket vegetation. HBL 20–26in; TL 1¼–1¾in; SH 13–16in; HL 1½–3½in; wt 6–9lb. Coat: thick, back gray and speckled, flanks variable gray to reddish; legs and forehead and nose bright reddish; white ring round eye. Nose slightly elongated.

Günther's dikdik
Madoqua güntheri

N Uganda, eastwards through Kenya and Ethiopia to the Ogaden and Somalia. Semi-arid scrub. HBL 24–29in; TL 1¼–2in; SH 13–15in; HL 1½–3½in; wt 9–12lb. Coat: back and flanks speckled gray, reddish nose and forehead. Nose conspicuously elongated.

Kirk's dikdik
Madoqua kirkii
Kirk's or Damaraland dikdik.

Tanzania and southern half of Kenya; Namibia and Angola. The two ranges are separate. HBL 24–28in; TL 1½–2½in; SH 14–17in; HL 1½–3in; wt 10–13lb. Coat: whitish ring round eye. Nose moderately elongated.

Genus *Oreotragus*

A single species.

Klipspringer
Oreotragus oreotragus

From the Cape of Angola, and up the eastern half of Africa to Ethiopia and E Sudan. Also two isolated massifs in Nigeria and Central African Republic. Well-drained rocky outcrops. Gait a stilted bouncing motion. Seldom solitary; occur in small family groups. Aggregations occur at favourable feeding sites. Lack pedal glands. HBL 29–35in; TL 2½–4in; SH 17–20in; HL 2–6in; wt 22–33lb. Coat: yellowish, speckled with gray, ears round, conspicuously bordered with black; black ring above hooves. Tail minute. Back conspicuously arched. Fur is thick, coarse and brittle, loosely rooted and lightweight, giving the animal a stocky appearance, hooves peg-like. Horns smooth, nearly vertical.

Genus *Raphicerus*

Coat reddish, large white-lined ears, tail minute, horns smooth and vertical. Three species.

Steenbuck
Raphicerus campestris

From Angola, Zambia and Mozambique southwards to the Cape, and in Kenya and Tanzania. Open, lightly wooded plains. Usually seen singly. Have large home ranges. Preorbital gland apparently not used for territorial marking. HBL 27–37in; TL 1½–2in; SH 18–24in; HL 3–7in; wt 22–33lb. Coat: reddish fawn; large white lined ears, shiny black nose. Horns smooth and vertical. Tail minute.

Sharpe's grysbuck
Raphicerus sharpei

Tanzania, Zambia, Mozambique, Zimbabwe. Woodland with low thicket or secondary growth. Mainly nocturnal and cryptic. HBL 24–29in; TL 2–3in; SH 18–24in; HL 2¼–4in; wt 16–25lb. Coat; Reddish brown, speckled with white on back and flanks; large white-lined ears. Horns small, smooth and vertical. Tail small. Back slightly arched.

Cape grysbuck
Raphicerus melanotis

Restricted to the southern Cape. Not as red as Sharpe's grysbuck, otherwise same.

Genus *Ourebia*

A single species.

Oribi
Ourebia ourebi

Range patchy and extensive. Eastern half of southern Africa, Zambia, Angola and Zaire, and from Tanzania northwards to Ethiopia, and westwards to Senegal. Grassy plains with only light bush, near water. Live mostly in pairs or small family groups. Aggregations do occur at favorable feeding sites. Scent glands and marking very well developed by males. Territories large. Commonly run with stotting gait. Grazer. HBL 36–55in; TL 2–6in; SH 21–26in; HL 3–7in; wt 30–46lb. Coat: reddish fawn back and flanks contrasting conspicuously with white underparts; forehead and crown reddish brown; black glandular spot below ear. tail short with black tip. Ears large and narrow. Horns short, vertical, slightly annulated at base.

Genus *Dorcatragus*

A single species.

Beira [V]
Dorcatragus megalotis

Somalia and Ethiopia, bordering the Red Sea and the Gulf of Aden. Stony barren hills and mountains. Very rare and little known. HBL 27–33in; TL 5–8in; SH 20–25in; HL 3–5in; wt 33–57lb. Gazelle-like. Large for a neotragine. Gray, finely speckled on back and flanks. Distinct dark band on sides. Underparts yellowish. Ears very large. Tail long and white. Horns widely separated, curving slightly forward. Rubbery hooves adapted to their rocky habitat.

Tribe Antilopini

Gazelles

Slender body and long legs. Male larger than female. Fawn upperparts, white underparts, typically with gazelline facial markings of dark band on blaze, white band on either side, dark band from eye to muzzle and white round eye. Horns generally S-shaped, annulated. Usually both sexes have horns, though female horns are shorter and thinner. Tail black tipped. Populations comprise typically territorial males, at least during the breeding season, female groups with their offspring, and bachelor goups of non-territorial males. Will migrate in response to seasonal changes in vegetation and climate, forming large mixed-sex aggregations during the winter or dry seasons. In seasonal parts of their range, birth peaks occur to coincide with vegetation flush in spring or early rains. Mixed feeders, but mostly browse.

Genus *Gazella*

Subgenus *Nanger*

Large gazelles, white rump and buttocks, tail white with black tip. 3 species.

Dama gazelle
Gazella dama

Sahara, from Mauritania to Sudan. Desert. Occur singly or in small groups. In rainy season move north into Sahara, in dry season back to Sudan. Very rare and disappearing. HBL 57–67in; TL 8–12in; SH 34–42in; HL 13–16in; wt 88–187lb. Long neck and legs for a gazelle. Coat: neck and underparts reddish brown, sharply contrasting with white rump, underparts and head; white spot on neck. Horns sharply curved back at base, relatively short.

Soemmerring's gazelle
Gazella soemmerringi

Horn of Africa, northwards to Sudan. Bush and acacia steppe. Occurs in small groups of 5–20. HBL 48–59in; TL 8–11in; SH 31–34in; HL 15–23in; wt 66–121lb. Coat: Pale fawn on head, neck and underparts; facial markings very pronounced. Short neck, long head. Horns sharply curved inwards at tips.

Grant's gazelle
Gazella granti

Tanzania, Kenya, and parts of Ethiopia, Somalia and Sudan. From semidesert to open savanna. Live in small groups of up to 30. Preorbital gland not used. HBL 55–65in; TL 8–11in; SH 29–36in; HL 18–32in; wt 83–180lb. Coat: black pygal band. Heavily built. Horns long, variable according to race.

Subgenus *Gazella*
Small gazelles. White on underparts and buttocks not extending to rump. Gestation five and a half months. HBL 27–42in; TL 6–10in; SH 16–27in; HL 10–17in; wt 33–70lb. Except *G. thomsoni*, occur in small groups. 7 species.

Mountain gazelle
Gazella gazella

Arabian peninsula, Palestine. Extinct over much of its range. Semidesert and desert scrub in mountains and coastal foothills. Breed seasonally. Coat: upperparts fawn, pygal and flank bands distinct.

Dorcas gazelle [E]
Gazella dorcas
Dorcas gazelle or jebeer.

From Senegal to Morocco, and westwards through N Africa and Iran to India. Semi-desert plains. Coat: upperparts pale, pygal and flank bands indistinct.

Slender-horned gazelle [E]
Gazella leptoceros

Egypt eastwards into Algeria. Mountainous and sandy desert. Coat: upperparts very pale. Ears large. Hooves broadened. Horns long, only slightly curved.

Red-fronted gazelle
Gazella rufifrons

From Senegal in a narrow band running eastward to Sudan. Semidesert steppe. Coat: reddish upperparts, narrow black band on side with reddish shadow band below, contrasting with white underparts. Horns short, stout, only slightly curved.

Thomson's gazelle
Gazella thomsoni

Tanzania and Kenya, and an isolated population in southern Sudan. Open grassy plains. Very abundant. Occur in large herds of up to 200. Have aggregations of several thousand during migration. Mixed feeders, but predominantly grazers. Largest of the subgenus. Coat: upperparts bright fawn, broad conspicuous dark band on the side; well-pronounced facial markings. Horns long, only slightly curved; small and slender in females.

Speke's gazelle [I]
Gazella spekei

Horn of Africa. Bare stony steppe. Little known. Coat: upperparts pale fawn, broad dark side-band. Small swollen extensible protuberance on nose.

Edmi [E]
Gazella cuvieri

Morocco, N Algeria, Tunis. Semidesert steppe. Coat: upperparts dark gray-brown, dark side-band with shadow band below; facial markings pronounced.

Subgenus *Trachelocele*
A single species.

Goitered gazelle
Gazella subgutturosa

From Palestine and Arabia eastwards through Iran and Turkestan to E China. Semidesert and desert steppe. In Asia, form winter aggregations of several thousand at lower altitudes to avoid snow. In summer disperse; Females disperse further than males. Stocky body, relatively short legs. HBL 15–43; TL 5–7in; SH 20–25in; HL 12–18in; wt 64–92lb.

Coat: upperparts pale, pygal and side-bands indistinct; facial markings not pronounced, fading to white with age. Horns arise close together and curved in at tips. Female mostly hornless. Male larynx forms conspicuous swelling.

Genus *Antilope*
A single species.

Blackbuck [*]
Antilope cervicapra

Indian subcontinent. From semidesert to open woodland. HBL 39–59in; TL 4–7in; SH 24–33in; wt 55–99lb. Coat: adult males, upperparts and neck dark brown to black, contrasting with white chin, eye and underparts; immature males and females light fawn. Horns long and spirally twisted. Females hornless.

Genus *Procapra*
Gazelle-like. Pale fawn upperparts, white rump and buttocks. 3 species.

Tibetan gazelle
Procapra picticaudata

Most of Tibet. Plateau grassland and high-altitude barren steppe. HBL 36–41in; TL 1–4in; SH 21–25in; HL 11–16in; wt 44–77lb. Face glands and inguinal glands absent. Horns S-shaped, not curving in at tips.

Przewalski's gazelle
Procapra przewalskii

China, from Nan Shan and Kukunor to Ordos Plateau. Semidesert steppe. Same size as Tibetan gazelle. Horns curve in at tip.

Mongolian gazelle
Procapra gutturosa

Most of Mongolia and Inner Mongolia. Dry steppe and semidesert. HBL 43–58in; TL 2–5in; SH 12–18in; wt 62–88lb. Small preorbital glands and large inguinal glands present.

Genus *Antidorcas*
A single species.

Springbuck
Antidorcas marsupialis

Southern Africa west of the Drakensberg Mountains and northwards to Angola. Open arid plains. Very gregarious. Used to migrate in vast herds of tens of thousands. Have characteristic pronking gait in which white hairs on

back are erected. Mixed feeeders, taking predominantly grass. HBL 38–45in; TL 8–12in; SH 29–33in; HL 14–19in; wt 55–101lb. Coat: bright reddish fawn upperparts, dark side-band contrasting with white underparts; face white with dark band from eye to muzzle; buttocks and rump white, and line of white erectile hairs in fold of skin along lower back. Horns short, sharply curved in at tip.

Genus *Litocranius*
A single species.

Gerenuk
Litocranius walleri

Horn of Africa south to Tanzania. Desert to dry bush savanna. Occur singly or in small groups. Browse on tall bushes by standing on hindlegs. Feeds delicately on leaves and young shoots. Independent of water. Do not form migratory aggregations. HBL 55–65in; TL 9–14in; SH 31–41; HL 12–17in; wt 63–114lb. Coat: reddish brown upperparts, back distinctly dark, white around eye and underparts. Very long limbs and neck with small head and weak chin. Tail short with black tip. Horns stout at base, sharply curved forward at tips. Female hornless.

Genus *Ammodorcas*
A single species.

Dibatag [V]
Ammodorcas clarkei

Horn of Africa. Grassy plains and scrub desert. Occur singly or in small groups. Able to stand on hindlegs to browse. Hold tail erect in flight. Do not form migratory aggregations. HBL 60–66in; TL 12–14in; SH 31–55in; HL 10–13in; wt 51–70lb. Coat: upperparts dark reddish gray, contrasting with white underparts and buttocks; head with chestnut gazelline markings; tail long, thin and black. Long neck and limbs. Horns curve backwards at base then forwards at tip like a reedbuck. Females hornless.

BPO'R

The Graceful Gerenuk

Survival in dry country

The delicate beauty of the gerenuk, with its extremely long, slender legs and neck, long ears and lyre-shaped horns, raises the question of how this seemingly frail creature is adapted to hot and dry country.

The gerenuk seems to have followed an evolutionary path similar to that of the giraffe, the long neck enabling it to reach higher up to gather food than other species.

Apart from size, the gerenuk differs from the giraffe, and from most other bovids, in that it habitually extends its reach even higher up by rising onto its hindlegs. In fact, all ungulates can rear up on their hindlegs—they do so when mating—but the gerenuck is unique in its maneuverability, sometimes stepping sideways round a tree while continuing to feed in a near-vertical stance. Minor adaptations in the skeleton and muscles of limbs and vertebral column facilitate this behavior, which enables the gerenuk to occupy an ecological niche different to those of its near relatives, the gazelles.

The gerenuk eats almost exclusively leaves, as well as some flowers and fruits, of a great variety of trees and shrubs, but no grass. Over 80 plant species are eaten by gerenuks in Tsavo National Park, Kenya. Their diet varies substantially between rainy and dry seasons, as the availability of different plant species changes too. Some of the plants are evergreens, with thick, rather hard leaves, coated with a thick cuticle which prevents evaporation of water. On evergreens, gerenuks are highly selective browsers, sniffing over a plant carefully before plucking small, tender leaves and shoots with their highly mobile lips. These and the narrow, pointed muzzle also permit them to pick the delicate leaves of acacias from among forbidding thorns.

The main advantage that the gerenuk derives from its highly selective feeding behavior is that it ingests only the juiciest, most nutritious plant parts. Probably, this is why gerenuks are independent of free water; even in captivity they hardly ever drink. This allows them to inhabit dry country that many other bovids would find too inhospitable. The gerenuk also has a less complicated digestive tract and weaker molar teeth, with a smaller grinding surface, than some of its relatives, eg Thomson's gazelle, that eat the more fibrous grass.

However, these benefits also have their costs: in the dry areas inhabited by the gerenuk, suitable food items are widely dispersed, particularly in the dry season; gathering enough to satisfy both water and energy requirements means spending considerable time searching for them. This has

two main consequences. Firstly, the individual gerenuk ought to minimize energy expenditure, so as to optimize its overall energy budget. A gerenuk can do this partly by avoiding strenuous activities such as fighting, long-distance, seasonal traveling, and partly by reducing energy losses in inclement weather. One way in which gerenuks—and some other gazelles— achieve this is to lie down in response to rainfall and/or strong wind. This reduces the surface area exposed to heat loss.

Secondly, gerenuk populations cannot achieve high biomass densities, compared to other species of hoofed mammals. In Tsavo National Park, where gerenuks are locally quite common, they contribute less than 0.5 percent to the total hoofed mammal biomass. This is also reflected in their social organization. Gerenuks do not form herds, but rather small groups of 1–5 individuals, although occasional aggregations of up to

▲ **Lacking the horns** of the male, the long pointed ears of the female gerenuk are her most distinctive feature.

▶ **Stilt walker.** The gerenuk is unique among antelopes in its ability to maneuver when at full stretch on its hindlegs. Like the giraffe, this ability to browse at heights allows the gerenuk to occupy a distinctive niche.

▼ **Courtship in the gerenuk** involves four stages: (1) the female raises her nose in a defensive gesture, while the male displays by a sideways presentation of the head; (2) the male marks the female from behind on the thigh; (3) the male performs the foreleg kick, tapping the females hindlegs with his foreleg; (4) the male performs the flehmen or lip-curl test, sampling the female's urine to determine whether she is ready to mate.

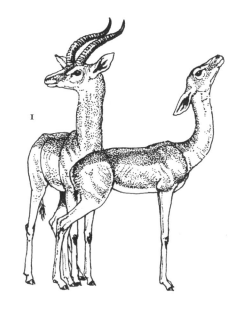

1

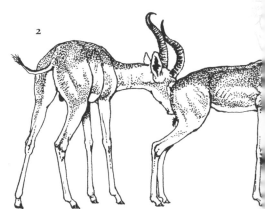

2

15 or 20 animals sometimes occur.

Even large groups rarely contain more than one adult male, as these are strictly territorial, excluding other males from their areas. Exclusion need not necessarily mean physical combat. Rather, the territorial system is maintained through frequent marking of dry twigs etc with the secretion of the scent glands in front of the eyes. Potential intruders apparently "respect" such marks or, at least, are able to assess the risk of progressing further. Such scent marking may be particularly important for male gerenuks, as their territories are comparatively large—0.8–1.6sq mi (2–4sq km).

Whereas other adult males are kept at a distance by the scent marks, a territorial male may tolerate subadult males near himself, as long as they acknowledge his superiority. Young males generally accompany their mothers before—at 1–1½ years—they start seeking the company of peers with which they form small all-male groups. Eventually, these break up, as each male begins to look for a territory of his own. Associations between females and adult, territorial males are usually short-lived, unless both reside in the same small area.

Overall, the gerenuk, although a moderately large antelope, shows a number of traits more typical of small, forest-living bovids, such as highly selective feeding, sedentary life, occupancy of a resource-type territory coupled with extensive scent marking, a rather weak tendency to form groups, and partial reliance on "freezing" to avoid predators. These characteristics enable the gerenuk to live, and even thrive, in a habitat that would appear to be only marginal for a herbivore of this size.   WL

GOAT ANTELOPES

Subfamily: Caprinae
Twenty-six species in 13 genera.
Order: Artiodactyla.
Distribution: Asia, C Europe, N and C America,
N Africa.

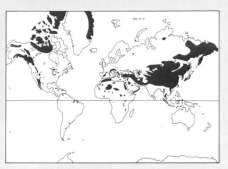

Habitat: steep terrain from hot deserts and most
jungles to arctic barrens.

Saiga
Tribe: Saigini.
Two species: **saiga** (*Saiga tatarica*), **chiru**
(*Pantholops hodgsoni*).

Rupicaprids
Tribe: Rupicaprini.
Five species in 4 genera.
Species include: **chamois** (*Rupicapra rupicapra*)
goral (*Nemorhaedus goral*), **Mainland serow**
(*Capricornis sumatrensis*), **Mountain goat**
(*Oreamnos americanus*).

Musk ox
Tribe: Ovibonini.
Two species: **Musk ox** (*Ovibos moschatus*),
takin (*Budorcas taxicolor*).

Caprids
Tribe: Caprini.
Seventeen species in 5 genera.
Species include: **argalis** (*Ovis ammon*), **Barbary
sheep** (*Ammotragus lervia*), **Blue sheep** (*Pseudois
nayaur*), **Himalayan tahr** (*Hemitragus
jemlahicus*), **ibex** (*Capra ibex*), **mouflon** (*Ovis
musimon*).

▶ **Species of goat antelopes** in order (left to
right) of increasing body size as a response to
increasing climatic severity. (1) Goral
(*Nemorhaedus goral*). (2) Himalayan tahr
(*Hemitragus jemlahicus*). (3) Urial (*Ovis
orientalis*). (4) Barbary sheep (*Ammotragus
lervia*). (5) Wild goat (*Capra aegagrus*).
(6) Japanese serow (*Capricornis crispus*).
(7) Chamois (*Rupicapra rupicapra*). (8) Ibex
(*Capra ibex*). (9) Mountain goat (*Oreamnos
americanus*). (10) Argalis (*Ovis ammon*). (11)
Takin (*Budorcas taxicolor*). (12) Musk ox (*Ovibos
moschatus*).

THE goat antelopes and their descendants (subfamily Caprinae) blossomed during the ice ages into a great diversity of species. They occupied extreme terrestrial environments such as hot deserts, arctic barrens, alpine plateaux, the steep, snow-covered cliffs at the edge of huge glaciers, and they survived in their evolutionary home, the humid tropics.

The basic goat-antelope body plan dates back to the mid-Tertiary period (35 million years ago), and is still reflected today in the serow of Southeast Asia. As the goat antelopes advanced into extreme climates, they abandoned the ancestral body plan, grew in size and diverged in appearance. The end products are giants such as the Musk ox from the arctic barrens, the giant rams from the highlands of central Asia, the long-horned ibex from the high alpine, and the bull-headed takin from the canyons of western China.

The goat antelopes and their kin are usually stocky, gregarious bovids that live on steep terrain. The primitive forms tend to have small horns and strongly patterned haircoats; the true sheep and goats are characterized by long, massive curving horns; the Musk oxen and their relatives are all large bodied with large horns. There is no consistent explanation for the very diverse body colors among these species, which vary from pure white, as in mountain goats and Dall's sheep, to pure black, as in the Musk oxen, serow or Stone's sheep. Gregarious forms tend to have strong frontal and rear markings. The females invariably are smaller in size and in the sheep and goats also have small horns.

At one extreme (the most primitive), there are so-called resource defenders, which live in small areas of highly productive and diverse habitats that supply all their needs

year round and can be easily defended against members of the same species.

At the opposite extreme (the most recently evolved) are the grazers, which live in less productive and climatically more severe habitats. They are highly gregarious, roam over larger areas, and have retained very few attributes of their tropical ancestors. They carry large horns, engage in a variety of "sporting engagements" and the sexes may be strikingly different.

In goat antelopes we find not only species adapted to each of these extremes, but also to a range of habitats in between—the evolutionary phases can be not only predicted but also observed. In broad terms, rupicaprids are resource defenders and their descendants are grassland exploiters.

Resource defense imposes a number of demands on life-style. Such animals usually lead a solitary life and are territorial. They defend their small patches of food resources against other members of their own species. To defend the resources they have weapons capable of inflicting the greatest damage to their opponents, such as short, sharp, piercing horns. Most are dark in coloration, with male and female alike in armament and appearance. Since it is easier to defend a small than a large territory, a territory holder must make maximum use of resources available; so such species are small bodied with selective food habits.

The little that is known of the serow fits with this picture of a resource defender. Male and female look alike; both have short and sharp horns and large preorbital glands for scent marking. Serow are aggressive and attempt to fight off predators, even Asian black bears. Their teeth are typical of browsers, with lower crowns than in other rupicaprids. Their legs are less specialized for climbing than are those of their advanced

relatives. Serow are spectacular with long, tasseled ears, a long neck-mane and dark color. Their home ranges are well structured, with trails, "horn-rubs" on saplings and dung heaps. Serow are widely distributed throughout tropical and subtropical Southeast Asia and regionally vary greatly in color. They are largely confined to moist, shrubby or timber-covered rock outcrops, but have penetrated cold temperate zones, giving rise to the Japanese serow, which can tolerate heavy snowfall. There are also small island species, such as the Taiwan serow. The mountain goat may be a serow that evolved in a glacial region.

The two extreme forms of rupicaprids are the European chamois (see pp150–151) and the American mountain goat. Both are adapted to cold, snowy mountains; both have glands on the back of the head. But here similarities end. The mountain goat is a massive, slow rock climber which only reverts to resource defense under severe winter conditions. It clings closely to cliffs and can live successfully even in the coastal ranges of Alaska and British Columbia with their high snowfalls. The chamois is relatively small bodied, gregarious, with a strikingly marked face and a bright rump patch. Only large males may usurp pockets of rich habitat in summer so as to fatten for the strenuous rut. In mountain goats, it is the adult females with a kid at heel which dominate males. In chamois, adult females also dominate males, but only young ones. The chamois' smaller size may be linked to its superlative dodging and evasive maneuvers in fighting; the mountain goat also evades but males have thick skin, especially thick on their rumps, for protection during fights.

In the goat antelopes, the break with resource defense correlates with exploitation of grasslands in open landscapes. Here, individuals band together in anonymous, but cooperative herds as a means of reducing predation. The larger the herd, the more secure each individual. Not only can many share in the task of vigilance and thereby permit each an adequate time for grazing, but animals in the center of the herd are virtually secure from predators. With grazing comes freedom from sedentary existence and an increase in tooth size to deal with the abrasive forage. Since grassland is productive, and all of it is available for feeding, a high density of grazers per unit area is possible. Several lines of goat antelopes are adapted to grazing, leading to the development of the caprids (sheep and goats), the Musk ox and probably the takin.

Among caprids, the most primitive form, related to the ancestral rupicaprids, is the tahr. This is intermediate in form and behavior between caprids and rupicaprids. It is closely wedded to cliffs and has broad food habits. It inhabits low latitudes in the Arabian peninsula, Nilgiri hills of southern India and the Himalayas, and has a very wide geographic distribution. The subspecies are divided by huge distances and oceans, which indicates their great age.

The Barbary sheep or aoudad from North Africa exemplifies the next stage in evolution. "New designs" arise at the periphery of the ancestor's range and only when colonizing a new habitat over a large area. Although a new species adapts to the environment it colonizes, changes in its external appearance arise, initially, as a consequence of social selection during colonization. The scarcity of colonizers prompts a high reproduction and social competition and rapid selection for larger bodies, stouter weapons and better bluffing abilities.

As expected, the Barbary sheep has a bigger head and horns than the tahr, yet it still retains a great diversity of combat forms, although head-clashing is clearly prevalent. In external appearance the Barbary sheep is close to goats, but biochemically it is closer to sheep.

The subsequent evolution of sheep and goats can be deciphered only in its broadest outlines. Both genera moved into cold mountains, increasing in body and horn size. However, there were also opportunistic local radiations which detract from a picture of orderly altitudinal and latitudinal progression. There must have been many hybridizations between populations during the ebb and flow of population movements, and there must have been many extinctions.

Goats are specialized for cliffs; sheep inhabit the open, rolling dry lands close to cliffs. These differences arise from different ways of dealing with predators: sheep escape by running and by clumping into cohesive herds, while goats are more specialized in putting obstacles and terrain with insecure footing in the way of pursuing predators. In external appearance, sheep and goat males differ considerably but females do not. In goats, the males carry long, scimitar-shaped, knobby horns as well as a chin beard. In sheep, males lack a chin beard and have massive curling horns; primitive sheep may have long hair on the cheeks and on the neck. Goats retain a tail and never have a large rump patch; sheep have short tails and large rump patches. Male goats spray themselves with urine and may consequently be quite malodorous. Where sheep and goats occur together they occupy different habitats, but this is not so when one occurs without the other. In the absence of sheep, some goats evolved horns not unlike those of sheep and lost their strong body odor. In the absence of goats, sheep move to cliffs and become ibex-like in body shape, such as American-type sheep. In both genera, as horns increase, the hair coat and body colors become less showy. Goats and sheep reach ever higher altitudes and latitudes, growing more grotesque and larger, specializing in combat delivering ever harder blows. Sheep had at least two great radiations: the first generated the 54/52 chromosome sheep (mouflon and American bighorn sheep), the second the giant argalis with 56 chromosomes. The extreme geographic forms, the bighorns and argalis, are remarkably similar in social behavior and appearance.

Asiatic and American sheep differ in ways of dealing with predators. The American sheep are ibex-like and cling to the vicinity of cliffs. Compared to Asiatic sheep, American sheep have much lower reproductive rates, for they do not bear twins, but they live longer. As populations, they cannot respond to decimation by a rapid build-up of numbers. They have a gestation period about one month longer than Asiatic sheep.

The goats had less success in geographic expansion than sheep. They stuck closely to cliffs, but along them they penetrated south as far as the Ethiopian highlands, and they moved west into Europe. In mountains without adjacent dry, grassy foothills goats apparently were more successful than sheep. The geographic expansion of goats and sheep is such that where these are found together primitive goats are paired with primitive sheep, and advanced goats with

▲ ▶ **The sheep from the goats.** Two species inhabiting the Rocky Mountains: an American bighorn ram ABOVE and a Mountain goat RIGHT. There could hardly be a greater contrast between two species, yet they are relatives, occupy the same mountains, endure similar weather, eat much the same food and escape from the same predators. The Mountain goat's hair is white and long; the sheep's is dun and short. The goat's horns are tiny, black and sharp; the sheep's are huge, light-colored and blunt. Goat fights are short, violent, bloody and rare; sheep fights are very long, ritualistic, mostly harmless and frequent. Sheep bang heads; goats do not. These differences are related to their very different life-styles: goats are resource defenders, becoming ruthlessly territorial when food is scarce on the cliffs in deep snow, whereas sheep are migratory, moving seasonally across wide expanses of forest.

eradicated in Eurasia and survives only in the Arctic, North America and Greenland.

The takin is still a mystery animal, which, like the goats, has dispensed with a collection of odoriferous glands in favor of a general body odor. The takin's domain comprises steep cliffs covered in varying part with shrubs and trees and having a temperate climate. Like the overgrown rupicaprid it appears to be, it still clings to forest cover, but like an open-country form with gregarious tendencies it will also flock to large, open alpine meadows.

To early man, caprids were of little economic significance as game animals. In the Neolithic, some eight millennia ago, sheep and goats were domesticated in the Near East. These animals quickly rose to supreme importance. Their value lies in their ability to exploit land of low productivity which cannot be otherwise exploited by man, the ease with which these can be controlled with little manpower, and their numerous valuable products. Sheep and goats are raised for meat, wool, milk products and fur. Goats in particular are associated with the image of the poor man's cow, and are severely destructive of sparse vegetation in arid regions. Domestic sheep and goats arose in the same geographic region. The mouflon gave rise to Domestic sheep, as indicated by the similarity of their chromosomes; Domestic goats arose from the common Wild goat. Neither ibex nor advanced sheep contributed to the domestications, although such could have been possible; wild sheep and goats are quite easily habituated to humans and can even be tamed in the wild in the absence of hunting. Caprids are greatly renowned as game animals and the heads of the giant sheep and bighorns rate as the greatest trophies of Asia and North America. In modern times, caprids have had a mixed history. In America, bighorn sheep have not done well and can be maintained only by judicious care, limited hunting and artificial reintroduction into areas of former occupation. Desert bighorns in particular are susceptible to diseases of livestock and to disturbances. In Asia, due to the high reproductive rates of Old World wild sheep and goats, the situation is a happy one in some areas, but in others, such as the Himalayan region, it is desperate. Severe competition from domestic stock and overhunting has eliminated wild caprids from many areas. In Europe, the artificial spread of ibex is a great success, as is the management of mouflons and chamois. The ibex was rescued from the verge of extinction and, in Canada, so was the Musk ox. VG

advanced sheep. The most primitive form is probably the markhor. The most advanced form is the ibex. These are stocky with thick, massive, knobby horns and a more uniform coat color. Species most distant from Asia Minor differ most from the wild goat. However, the evolutionary pattern of goats is complex, with some species assuming sheep-like shapes and behavior. Farthest from the Middle East and most sheep-like of all goats is the Blue sheep of Tibet and western China.

The Caprini are one radiation from the Rupicaprini that adapted to open landscapes and temperate climates. The Musk oxen radiation is another one, as is probably that of the takin. Both evolved frontal combat and head banging. With that comes hierarchical social structures, male groups and a high degree of gregariousness. The interpretation of these groups is difficult due to extinctions. Only the most highly evolved of the Musk oxen remains, and even it was

THE 26 SPECIES OF GOAT ANTELOPES

Tribe Saigini

Formerly considered caprids but now normally classified as gazelles (subfamily Antilopinae). Like gazelles they have inflatable bags beside nostrils and carpal glands. Only males have horns. Inguinal glands are present.

Saiga
Saiga tatarica

N Caucasus, Kazakhstan, SW Mongolia, Sinkiang (China). Cold, arid steppe. Male HTL 48–57in; HT 27–31in; wt 70–112lb. Female HTL 42–49in; HT 22–29in; wt 46–90lb. Coat: sandy; horns amber and translucent. Head with grotesque proboscis that increases in size in males during rut, at which time hair tufts below eyes become impregnated with sticky smelly preorbital gland secretions; ears and tail short; neck with short mane. Twinning common. Population greatly expanded in recent years. Gestation: 145 days.

Chiru
Pantholops hodgsoni
Chiru or Tibetan antelope.

Tibet, Tsinghai, Sichuan (China), Ladak (N India). High Tibetan plateau. HTL 67in; HT 35–39in; wt up to 88lb in males. Coat: sandy, horns black, 20–28in; rump patch large; face of male dark. Tail and ears short; nasal sacs when inflated size of pigeon eggs; auxiliary glands present. Gestation: about 180 days.

Tribe Rupicaprini

The most primitive caprids. Defend resources using short sharp horns. Little difference between sexes in weight and appearance. Tail short. Strong affinity for steep slopes and shrub or tree cover. Food habits broad with emphasis on browse. Four teats. Young follow mother immediately after birth.

Mainland serow *
Capricornis sumatrensis

Tropical and subtropical E Asia. Male HTL unknown; HT up to 42in; wt up to 310lb. Female HTL unknown; HT up to 35in; wt up to 220lb. Coat: very variable, often dark, with long (16in) white or brown neck mane; hair coarse, long and bristly; short beard from mouth to ears. Horns in male stouter than in female, 6–10in; ears long, lance-shaped; large preorbital glands and foot glands.

Japanese serow
Capricornis crispus
Japanese or Taiwanese serow.

Japan, Taiwan. Particularly adapted to low temperatures and snowy climates in Japan. Dwarfed island form on Taiwan. Smaller and lighter colored than Mainland serow with conspicuous cheek beards in both sexes.

Goral *
Nemorhaedus goral
Goral, Red goral, Common goral.

N India and Burma to SE Siberia and south to Thailand. Very steep cliffs in dry climates; Manchurian form in moist, snowy climate. Male (Manchuria) HTL 41–46in; HT 27–31in; wt 62–92lb. Female HTL 42–46in; HT 20–29in; wt 48–77lb. Coat: highly variable, ranging from foxy red to dark gray to white; throat patch white; body hair long; neck mane short in male; horns black up to 9in in males, 8in in females. Tail longer than in serow, with terminal tuft; tail longer in Manchurian form than in southern races. Rudimentary preorbital glands. Gestation: 250–260 days in Manchurian form, shorter in southern forms. 8 races from India, China, E Siberia. Endangered in Far East.

Chamois
Rupicapra rupicapra

NW Spain, Italy, Pyrenees to Caucasus, Carpathian and Tatra Mts; NE Turkey; Balkans; introduced to New Zealand. Moist alpine and subalpine, particularly open mountain. HTL 49–53in; HT 27–31in; wt 66–110lb (male), 53–92lb (female). Coat: in alpine form, pale brown in summer, dark brown/black with white rump patch and white to yellow facial stripes in late fall/winter. In southwestern form, reddish beige in summer, dark brown with wide yellowish patches on throat, neck sides, shoulders, hindlimbs in late fall/winter. Occipital glands present in both sexes. Tail short. Gestation 160–170 days. 9 subspecies.

Mountain goat
Oreamnos americanus

SE Alaska, S Yukon and SW Mackenzie to Oregon, Idaho and Montana; introduced in other N American mountain areas. Steep cliffs and edge of major glaciers in areas of high snowfall. Size very variable regionally: male HTL up to 69in; HT up to 48in; wt up to 310lb. Female HTL up to 57in; HT up to 36in; wt up to 125lb. In southern forms, wt 154lb (male) and 126lb (female). Coat: yellowish-white, with long underwool and longer guard hairs that elongate into a stiff mane on neck and rump and into "pantaloons" on legs; molts in July; horns black, 6–10in long, thicker in males. Body with massive legs and very large hooves. Tail short. Gestation: about 180 days.

Tribe Ovibonini

Grotesque giant rupicaprids from arctic and alpine environments, with long-haired, massive squat bodies carried on short, stout legs. Sexes similar in appearance, but females only 60 percent of males' weight. Horns relatively larger than in rupicaprids, but smaller than in caprids, curved for frontal attacks. Single young; gestation 8–8.5 months. Tail short. Four teats.

Takin
Budorcas taxicolor
Takin or Golden-fleeced cow.

W China, Butan, Burma. High alpine and subalpine bamboo forests on steep rugged terrain. Size variable: Male HTL 67–87in; HT 39–51in; wt up to 770lb, female up to 550lb. Coat: normally light, long and shaggy, golden in one race; face dark in bulls, but only nose dark in cows and calves. No localized skin glands, but oily, strong smelling substance with burning taste, secreted over whole body.

Musk ox
Ovibos moschatus

Alaska to Greenland. Arctic and tundra near glaciers; adapted to extreme cold; mixed feeder, restricted to areas with thin snow. Size variable, increasing from north to south: Male HTL (south) 96in; HT 57in; wt 770lb; male (Greenland) HTL 71in; HT 43–51in; wt up to 1,435lb (in captivity). Coat: black with light saddle and front, possibly bleaching in spring, dense and long, with hair strands up to 24in. Foot glands present; preorbital glands large and secrete copiously in bull in rut.

Tribe Caprini

Advanced caprids with large differences in weight and external appearance between sexes. Specialized for frontal combat, with hierarchical social system in males. Spatial segregation of sexes outside the mating season. Horns with annual growth, much larger in males; length of display hair and horn mass inversely related. Teeth strongly molarized. Teats 2–4. Strong preference for open areas and grazing. Highly gregarious.

Genus *Hemitragus*

Most primitive and rupicaprid-like in form and behavior of the caprini. Both sexes horned. Those of males stouter, longer and heavier, up to 17in. Prefer steep, arid or cold cliffs. Males grow thick, heavy skin. Not malodorous, lack foot, groin, preorbital or tail glands. Females about 60 percent of male weight. Gestation: 180 days.

Himalayan tahr
Hemitragus jemlahicus

Himalayas. Introduced to New Zealand. HTL 51–67in; HT 24–39in; wt may exceed 238lb in males. Coat: variable over body from copper to black; hair relatively long with long-haired, shaggy neck ruff in males; haircoat shorter in females.

Nilgiri tahr V
Hemitragus hylocrius

S India. HT 24–39in. Coat: dark, almost black, short-haired, with silver saddle in males; mane in male short and bristly; females grayish brown, with a white belly.

Arabian tahr E
Hemitragus jayakari

Oman. HTL, HT, wt: not known; is said to be the smallest species. Coat: brownish with dark dorsal stripe, black legs, white belly and long hair about jaw and withers.

Genus *Ammotragus*
A single species.

Barbary sheep *
Ammotragus lervia
Barbary sheep or aoudad.

N Africa. Mountains, particularly high desert. Male HTL 61–65in; HT 35–39in; wt up to 310lb; females 50 percent of male wt. Coat: short with long-haired neck ruffs and pantaloons on front legs in both sexes; beard and rump patch absent. Tail long, flat, naked on underside. Callouses on knees. Subcaudal glands present. Males do not smell offensively. Horns in males up to 33in, in females up to 16in. Ears slim and pointed. Gestation period 160 days. 4 subspecies.

Genus *Pseudois*
A single species.

Blue sheep
Pseudois nayaur
Blue sheep or bharal.

Himalayas, Tibet, E China. Slopes close to cliffs. Few measurements available: male ht 36in; wt 132lb; females wt 88lb. Coat: bluish with light abdomen, legs strongly marked; lacks beard and shin glands; rump patch small; tail naked on underside. No callouses on knees. Horns curve rearward, cylindrical, up to 33in long in males, tiny in females. Ears pointed and short. Gestation: 160 days. Longevity: average 9 years, but up to 24 years.

Genus *Capra*
Males with chin beards and strong smelling. No preorbital, groin, or foot glands; anal glands present. Both sexes have callouses on knees and long, flat tails with bare underside. Ears long, pointed, except in alpine forms. Rump patch small. Cliff-adapted jumpers, highly gregarious, which penetrate alpine and desert environments. Body size varies locally. Females 50–60 percent of male weight. In males, horns increase in length and weight with age. Female horns 8–10in in all species. Breeding coat of male becomes more colorful with age in most species. Gestation: 150 days in small-bodied species, up to 170 days in large-bodied species. Twins common. Average life expectancy about 8 years. Many races, exact number unsettled.

Wild goat
Capra aegagrus
Wild goat or bezoar.

Greek Islands, Turkey, Iran, SW Afghanistan, Oman, Caucasus, Turkmenia, Pakistan and adjacent India; Domestic goat worldwide. Size highly variable: male htl (Crete) unknown; ht unknown; wt 57–92lb; male (Persia) up to 198lb, female up to 99lb. Coat: old males very colorful compared to females or young males. Horns flat and scimitar-shaped, 23–50in. 4 subspecies, including the Domestic goat (*C. a. hircus*).

Spanish goat
Capra pyrenaica
Spanish goat or Spanish ibex.

Pyrenees Mts. Male htl 51–55in; ht 25–27in; wt 143–176lb. Female htl 39–43in; ht 27–29in; wt 77–99lb. Coat: similar to Wild goat in color. Horns differ from those of ibex or goats, up to 30in. 4 subspecies.

Ibex
Capra ibex
C Europe, Afghanistan and Kashmir to Mongolia and C China, N Ethiopia to Syria and Arabia. Extreme alpine of desert. Male htl (Siberian) 45–67in; ht 25–41in; wt 176–220lb. Female htl 57in; ht 25–27in; wt 66–110lb. Alpine ibex may be larger than this (wt up to 257lb); Nubian and Walia ibex are smaller. Coat: less colorful than in Wild goat; uniformly brown in alpine races; chin beards smaller than in goats. Horns massive and thick, but much more slender than in Caucasian turs: maximum length 33in (Alpine), 50–56in (Siberian).

Markhor [v] [*]
Capra falconeri
Afghanistan, N Pakistan, N India, Kashmir, S Uzbekistan, Tadzhistan. Woodlands low on mountain slopes. Male htl 63–66in; ht 34–39in; wt 176–242lb. Female htl 55–59in; ht 25–27in; wt 70–88lb. Coat: diagnostic due to long neck mane in male, pantaloons and strong markings; female does not have display hairs. Horns twisted, maximum length 32–56in. The largest goat. Dentition more primitive than ibex and turs.

East Caucasian tur
Capra cylindricornis
E Caucasus Mts. Male htl 57–59in; ht 31–38in; wt 143–220lb; Female htl 47–55in; ht 25–27in; wt 99–121lb. Coat: uniformly dark brown in winter, red in summer; chin beard very short, up to 3in. Horns cylindrical and sharply backward winding, as in *Pseudois*; maximum length 40in.

West Caucasian tur
Capra caucasica
W Caucasus Mts. Size and form as East Caucasian tur except that horns are similar to those of alpine ibex, but more massive and curved. Skull form diagnostic and different from ibex. Chin beard up to 7in.

Genus *Ovis*
Characterized by the presence of preorbital, foot and groin glands. Males do not have offensive smell. Tail short; rump patch small in primitive, large in advanced species. Females 60–70 percent of male weight; two teats. Horns present in both sexes, except a few mouflon populations where females lack horns; female horns normally very small. Horns increase in mass from urials to argalis sheep. In the latter horns form up to 13 percent of the male's body mass. In mouflons the horns may wind backwards, otherwise horns wind forwards. Horn mass in large mature rams; Urials 11–20lb; Altai argalis 44–48lb, exceptionally more; Snow sheep about 66–110lb; Stone's and Dall's sheep 18–22lb; bighorns rarely in excess of 26lb. Horns used as sledge hammers in combat. All species prefer grazing. Highly gregarious, with sexes segregated except at mating season.

Urial [*]
Ovis orientalis
Kashmir to Iran, particularly rolling terrain and deserts. Male htl 43–57in; ht 35–39in; wt 79–192lb. Coat: color variable, usually light brown; males with whitish cheek beards and light-colored long neck ruff; rump patch diffuse; tail thin and long for a sheep. Horns relatively light, forward winding. Long-legged, fleet-footed. Gestation: 150–160 days. Longevity: 6 years.

Argalis [*]
Ovis ammon
Pamir to Outer Mongolia and throughout Tibetan plateau. Cold, high alpine and cold desert habitats. Male htl 71–79in; ht 43–49; wt 210–310lb (400lb in Altai argalis); largest sheep. Coat: light brown with large white rump patch and white legs; size of neck ruff inversely related to size of horns; horns up to 75in long and 20in in circumference. Twinning common.

Mouflon
Ovis musimon
Asia Minor, Iran, Sardinia, Corsica, Cyprus; widely introduced in Europe. Cold and desert habitats. Male htl 43–51in; ht 25–29in; wt 55–121lb. Coat: dark chestnut brown with light saddle; rump patch distinct; lacks a cheek bib but possesses dark ruff; tail short, broad and dark. Face of adults becomes lighter with age. Smallest wild sheep.

Snow sheep
Ovis nivicola
Snow sheep or Siberian bighorn.

NE Siberia. Extreme alpine and arctic regions, particularly cliffs. Male htl 64–70; ht 35–39in; wt 198–265lb; females 132–143lb. Coat: dark brown with small, distinct rump patch and broad tail; some races light colored. 4 races.

Thinhorn sheep
Ovis dalli
Thinhorn, Stone's, Dall's or White sheep.

Alaska to N British Columbia. Extreme alpine and arctic regions, particularly cliffs. Male htl 53–61in; ht 36–40in; wt 198–265lb. Coat: white in 2 subspecies, black or gray in the Stone's sheep; large, distinct rump patch in latter.

American bighorn sheep [*]
Ovis canadensis
American bighorn sheep or Mountain sheep.

SW Canada to W USA and N Mexico. Alpine to dry desert, particularly cliffs. Male htl 66–73in; ht 37–43in; wt 125–310lb depending on locality; female 123–176lb. Coat: light to dark brown, lacks ruff or cheek beards; with large rump patches. Body stocky as in ibex. Gestation: 175 days. Longevity: average 9 years but to 24 years. 7 races.

VG

The Nimble Chamois

Social life of a mountain goat

The extraordinary agility of the chamois seems to defy gravity. Only a few hours after birth, a kid can nimbly follow its mother along narrow ledges or down precipitous screes. This agility enables them to survive predation from the eagles, wolves, lynxes and Brown bears that share their range.

Chamois live in mountainous country in Europe from the Cantabrians to the Caucasus and from the High Tatra to the Central Apennines. There are 10 subspecies, divided into two groups: the Southwestern races and the Alpine races. Some recent evidence suggests that these two groups should be recognized as separate species. In any case, the behavioral ecology of different chamois populations varies greatly and the species' social system is very flexible.

Some 400 of the Apennine chamois survive, all on a few mountains of the Abruzzo National Park, Central Italy, where they are strictly protected. The females with young live mainly in woodland during the cold months, but they move to Alpine meadows in late spring until the end of the fall.

Females are usually resident and tend to live in flocks, the largest group sizes being reached in late summer, prior to the rut. Males stay with the mother's flock until sexual maturity at 2–3 years old. Then, they live nomadically until they reach full maturity at 8–9 years old, when they become attached to a definite area. This may or may not lie within the females' range, and so not all males have access to females during the rut. However, should a female emigrate

because of a high population density, she is likely to meet one such peripheral male and be fertilized.

Fully adult males live a solitary life in steep, rocky woodland throughout most of the year, only joining the female flocks occasionally, whereas young males mix readily with these. During the rut, the young males are chased away by older males, which defend their harems and begin spending more and more time with them on the Alpine meadows. The peak of the rut is reached in mid-November. Harem holders have to chase rivals off the harem, to test females for receptivity and prevent them from leaving: difficult, energy-consuming tasks which the harem holder has to undertake just before the winter. Despite the harem male's efforts, the females become skittish and restless when they come into heat, and the harem scatters. Thus, peripheral males have chances to mate too. This makes the advantage of being a harem-holder difficult to understand; it may lie in saving time, since courtship of a receptive female may last for hours, but a familiar male is likely to be accepted earlier than a stranger.

With the first heavy snowfalls, the chamois flocks split up and move to the woodland winter ranges. There they feed upon sparse, scattered food sources (buds, lichens, small grass patches) which would make living in a flock disadvantageous through increasing food competition. Heavy winter mortality may be suffered, especially

▲ **Winter quarters.** Chamois at 6,900ft (2,100m) in the Swiss Alps.

▶ **Out on the edge.** Chamois leaping down a rockface in the Abruzzo National Park, Italy.

◀ **Gestures of the chamois.** (1) "Side display" is an intimidatory stance used mainly by females and young males: the head is held high, the back is arched and the animal moves stiffly towards the receptor, which may move away in an alarmed "tail-up" posture (2). Grown-up males prefer to use a comparable behavior pattern where the neck stretches upwards and the long hair on withers and hind quarters is erected (3). The tail sticks between the rumps. Thus, an apparent increase of body size is achieved. Often subordinates creep away from displaying dominants in an inconspicuous, submissive, "low-stretching" attitude (4).

by the youngest age classes in relation to the duration of the snow cover. Also, the nomadic habits of young males may lead them to ecologically unsuitable areas.

Kids are born in May–June on rocky, secluded, inaccessible areas, where pregnant females have isolated themselves just before delivering. Large groups do not form again until mid-June. Normally, a mature female will give birth to a single kid per year.

Chamois are aggressive among themselves to the extent that potentially lethal fights are common. Normally, however, a vast repertoire of vocal and visual threat displays lessens the danger of direct forms of aggressiveness. In a fight, chamois show no inhibitions about goring rivals, hooking them in the throat, chest or abdomen, unless the loser is quick to lie flat on the ground, stretching forward its neck in an extreme submissive posture resembling an exaggeration of the suckling posture. Probably, infantile mimicry works to soothe the attacker's aggressiveness, but it is also the only posture which prevents the latter from goring efficiently; in fact the chamois horns are strongly crooked at their tops, so that the blows must be delivered with an upward motion to be effective. By lying down the loser simply prevents the winner from using its weapons!   SL

BIBLIOGRAPHY

The following list of titles indicates key reference works used in the preparation of this volume and those recommended for further reading.
The list is divided into two categories, those books on mammals in general and those specifically devoted to hoofed mammals.

General

Boyle, C. L. (ed) (1981) *The RSPCA Book of British Mammals*, Collins, London.

Corbet, G. B. and Hill, J. E. (1980) *A World List of Mammalian Species*, British Museum and Cornell University Press, London and Ithaca, N. Y.

Dorst, J. and Dandelot, P. (1972) *Larger Mammals of Africa*, Collins, London.

Grzimek, B. (ed) (1972) *Grzimek's Animal Life Encyclopedia*, Vols 10, 11 and 12, Van Nostrand Reinhold, New York.
Hall, E. R. and Kelson, K. R. (1959) *The Mammals of North America*, Ronald Press, New York.

Harrison Matthews, L. (1969) *The Life of Mammals*, vols 1 and 2, Weidenfeld & Nicolson, London.

Honacki, J. H., Kinman, K, E. and Koeppl, J. W. (eds) (1982) *Mammal Species of the World*, Allen Press and Association of Systematics Collections, Lawrence, Kansas.

Kingdon, J. (1971–82) *East African Mammals*, vols I–III, Academic Press, New York.

Morris, D. (1965) *The Mammals*, Hodder & Stoughton, London.

Nowak, R. M. and Paradiso, J. L. (eds) (1983) *Walker's Mammals of the World* (4th edn), 2 vols, Johns Hopkins University Press, Baltimore and London.

Vaughan, T. L. (1972) *Mammalogy*, W. B. Saunders, London and Philadelphia.

Young, J. Z. (1975) *The Life of Mammals: their Anatomy and Physiology*, Oxford University Press, Oxford.

Hoofed Mammals

Chaplin, R. E. (1977) *Deer*, Blandford, Poole, Dorset, England.

Chapman, D. and Chapman, N. (1975) *Fallow Deer – Their History, Distribution and Biology*, Terrence Dalton, Lavenham, Suffolk, England.

Clutton-Brock, T. H., Guinness, F. E. and Albon, S. D. (1982) *Red Deer – Behaviour and Ecology of Two Sexes*, Edinburgh University Press, Edinburgh.

Dagg, A. I. and Foster, J. B. (1976) *The Giraffe – its Biology, Behavior and Ecology*, Van Nostrand Reinhold, New York.

Eltringham, S. K. (1982) *Elephants*, Blandford, Poole, Dorset, England.

Gauthier-Pilters, H. and Dagg, A. I. (1981) *The Camel – its Evolution, Ecology, Behavior and Relationship to Man*, University of Chicago Press, Chicago.

Geist, V. (1971) *Mountain Sheep – a Study in Behavior and Evolution*, University of Chicago Press, Chicago.

Groves, C. P. (1974) *Horses, Asses and Zebras in the Wild*. David and Charles, Newton Abbot, England.

Haltenorth, T. and Diller, H. (1980) *A Field Guide to the Mammals of Africa Including Madagascar*, Collins, London.

Kingdon, J. (1979) *East African Mammals*, vol III, parts B, C, D, Academic Press, London and New York.

Laws, R. M., Parker, I. S. C. and Johnstone, R. C. B. (1975) *Elephants and Their Habitats – the Ecology of Elephants in North Bunyoro, Uganda*, Clarendon Press, Oxford.

Leuthold, W. (1977) *African Ungulates – a Comparative Review of their Ethology and Behavioral Ecology*, Springer-Verlag, Berlin.

Mloszewski, M. J. (1983) *The Behavior and Ecology of the African Buffalo*, Cambridge University Press, Cambridge.

Moss, C. (1976) *Portraits in the Wild – Animal Behaviour in East Africa*, Hamish Hamilton, London.

Nievergelt, B. (1981) *Ibexes in an African Environment – Ecology and Social System of the Walia Ibex in the Simen Mountains, Ethiopia*, Springer-Verlag, Berlin.

Schaller, G. B. (1967) *The Deer and the Tiger – a Study of Wildlife in India*, University of Chicago Press, Chicago.

Schaller, G. B. (1977) *Mountain Monarchs – Wild Sheep and Goats of the Himalaya*, University of Chicago Press, Chicago.

Sinclair, A. R. E. (1977) *The African Buffalo*, University of Chicago Press, Chicago.

Spinage, C. A. (1982) *A Territorial Antelope – The Uganda Waterbuck*, Academic Press, London.

Walther, F. R., Mungall, E. C. and Grau, G. A. (1983) *Gazelles and their Relatives – A Study in Territorial Behavior*, Noyes Publications, Park Ridge, New Jersey.

Whitehead, G. K. (1972) *Deer of the World*, Constable, London.

GLOSSARY

Abomasum   the final chamber of the four sections of the RUMINANT ARTIODACTYL stomach (following the RUMEN, RETICULUM and OMASUM). The abomasum alone corresponds to the stomach "proper" of other mammals and the other three are elaborations of its proximal part.

Adaptive radiation   the pattern in which different species develop from a common ancestor (as distinct from CONVERGENT EVOLUTION, a process whereby species from different origins became similar in response to the same SELECTIVE PRESSURES).

Adult   a fully developed and mature individual, capable of breeding, but not necessarily doing so until social and/or ecological conditions allow.

Allantoic stalk   a sac-like outgrowth of the hinder part of the gut of the mammalian fetus, containing a rich network of blood vessels. It connects fetal circulation with the PLACENTA, facilitating nutrition of the young, respiration and excretion. (See CHORIOALLANTOIC PLACENTATION.)

Allopatry   condition in which populations of different species are geographically separated (cf SYMPATRY).

Alpine   of the Alps or any lofty mountains; usually pertaining to altitudes above 1,500m (4,900ft).

Amphibious   able to live both on land and in water.

Amynodont   a member of the family Amynodontidae, large rhinoceros-like mammals (order Perissodactyla), which became extinct in the Tertiary.

Anal gland or sac   a gland opening by a short duct either just inside the anus or on either side of it.

Ancestral stock   a group of animals, usually showing primitive characteristics, which is believed to have given rise to later, more specialized forms.

Anthracothere   a member of the family Anthracotheriidae (order Artiodactyla), which became extinct in the late Tertiary.

Antigen   a substance, whether organic or inorganic, that stimulates the production of antibodies when introduced into the body.

Antrum   a cavity in the body, especially one in the upper jaw bone.

Aquatic   living chiefly in water.

Arboreal   living in trees.

Artiodactyl   a member of the order Artiodactyla, the even-toed ungulates.

Astragalus   a bone in the ungulate tarsus (ankle) which (due to reorganization of ankle bones following reduction in the number of digits) bears most of the body weight (a task shared by the CALCANEUM bone in most other mammals).

Baculum (os penis or penis bone)   an elongate bone present in the penis of certain mammals.

Bifid (of the penis)   the head divided into two parts by a deep cleft.

Biotic community   a naturally occurring group of plants and animals in the same environment.

Blastocyst   see IMPLANTATION.

Bovid   a member of the cow-like artiodactyl family, Bovidae.

Brachydont   a type of short-crowned teeth whose growth ceases when full-grown, whereupon the pulp cavity in the root closes. Typical of most mammals, but contrast the HYPSODONT teeth of many herbivores.

Brindled   having inconspicuous dark streaks or flecks on a gray or tawny background.

Brontothere   a member of the family Brontotheriidae (order Perissodactyla), which became extinct in the early Tertiary.

Browser   a herbivore which feeds on shoots and leaves of trees, shrubs etc, as distinct from grasses (cf GRAZER).

Bullae (auditory)   globular, bony capsules housing the middle and inner ear structures, situated on the underside of the skull.

Bunodont   molar teeth whose cusps form separate, rounded hillocks which crush and grind.

Calcaneum   one of the tarsal (ankle) bones which forms the heel and in many mammalian orders bears the body weight together with the ASTRAGALUS.

Camelid   a member of the camel family, Camelidae, of the Artiodactyla.

Cameloid   one of the South American camels.

Cannon bone   a bone formed by the fusion of METATARSAL bones in the feet of some families.

Caprid   a member of the tribe Caprini of the Artiodactyla.

Carnivore   any meat-eating organism (alternatively, a member of the order Carnivora, many of whose members are carnivors).

Caudal gland   an enlarged skin gland associated with the root of the tail. Subcaudal: placed below the root; supracaudal: above the root.

Cecum   a blind sac in the digestive tract, opening out from the junction between the small and large intestines. In herbivorous mammals it is often very large; it is the site of bacterial action on cellulose. The end of the cecum is the appendix; in species with reduced ceca the appendix may retain an antibacterial function.

Cellulose   the fundamental constituent of the cell walls of all green plants, and some algae and fungi. It is very tough and fibrous, and can be digested only by the intestinal flora in mammalian guts.

Cementum   hard material which coats the roots of mammalian teeth. In some species, cementum is laid down in annual layers which, under a microscope, can be counted to estimate the age of individuals.

Cervid   a member of the deer family (Cervidae), of the Artiodactyla.

Cervix   the neck of the womb (cervical—pertaining to neck).

Chalicothere   a member of the family Chalicotheriidae (order Perissodactyla), which became extinct in the Pleistocene.

Cheek-teeth   teeth lying behind the canines in mammals, comprising premolars and molars.

Chorioallantoic placentation   a system whereby fetal mammals are nourished by the blood supply of the mother. The chorion is a superficial layer enclosing all the embryonic structures of the fetus and is in close contact with the maternal blood supply at the placenta. The union of the chorion (with its vascularized ALLANTOIC STALK and YOLK SAC) with the placenta facilitates the exchange of food substances and gases, and hence the nutrition of the growing fetus.

Class   taxonomic category subordinate to a phylum and superior to an order.

Cloaca   terminal part of the gut into which the reproductive and urinary ducts open. There is one opening to the body, the cloacal aperture, instead of a separate anus and urinogenital opening.

Cloud forest   moist, high-altitude forest characterized by dense UNDERSTORY growth, and abundance of ferns, mosses, orchids and other plants on the trunks and branches of the trees.

Colon   the large intestine of vertebrates, excluding the terminal rectum. It is concerned with the absorption of water from feces.

Concentrate selector   a herbivore which feeds on those plant parts (such as shoots and fruits) which are rich in nutrients.

Conspecific   members of the same species.

Conspecific   member of the same species.

Convergent evolution   the independent acquisition of similar characters in evolution, as opposed to possession of similarities by virtue of descent from a common ancestor.

Crepuscular   active in twilight.

Crypsis   an aspect of the appearance of an organism which camouflages it from the view of others, such as predators or competitors.

Cryptic (coloration or locomotion)   protecting through concealment.

Cue   a signal, or stimulus (eg olfactory) produced by an individual which elicits a response in other individuals.

Cursorial   being adapted for running.

Cusp   a prominence on a cheek-tooth (premolars or molar).

Delayed implantation   see IMPLANTATION.

Dental formula   a convention for summarizing the dental arrangement whereby the numbers of each type of tooth in each half of the upper and lower jaw are given; the numbers are always presented in the order: incisor (I), canine (C), premolar (P), molar (M). The final figure is the total number of teeth to be found in the skull. A typical example for Carnivora would be I3/3, C1/1, P4/4, M3/3 = 44.

Dentition   the arrangement of teeth characteristic of a particular species.

Dermis   the layer of skin lying beneath the outer EPIDERMIS

Desert   areas of low rainfall, typically with sparce scrub or grassland vegetation or lacking vegetation altogether.

Dicerathere   a member of the family Diceratheriidae (order Perissodactyla), which became extinct in the Miocene.

Digit   a finger or toe.

Digital glands   glands that occur between or on the toes.

Digitgrade   method of walking on the toes without the heel touching the ground (cf PLANTIGRADE).

Diprotodont   having the incisors of the lower jaw reduced to one functional pair, as in possums and kangaroos (small, non-functional incisors may also be present). (cf POLYPROTODONT.)

Disjunct or **discontinuous distribution**   geographical distribution of a species that is marked by gaps. Commonly brought about by fragmentation of suitable habitat, especially as a result of human intervention.

Dispersal   the movements of animals, often as they reach maturity, away from their previous home range (equivalent to emigration). Distinct from dispersion, that is, the pattern in which things (perhaps animals, food supplies, nest sites) are distributed or scattered.

Display   any relatively conspicuous pattern of behavior that conveys specific information to others, usually to members of the same species; can involve visual and or vocal elements, as in threat, courtship or "greeting" displays.

Distal   far from the point of attachment or origin (eg tip of tail).

Diurnal   active in daytime.

Dorsal   on the upper or top side or surface (eg dorsal stripe).

Ecology   the study of plants and animals in relation to their natural environmental setting. Each species may be said to occupy a distinctive ecological NICHE.

Ecosystem   a unit of the environment within which living and nonliving elements interact.

Ecotype   a genetic variety within a single species, adapted for local ecological conditions.

Edentate   a member of an order comprising living and extinct anteaters, sloths, armadillos (XENARTHRANS), and extinct paleanodonts.

Elongate   relatively long (eg of canine teeth, longer than those of an ancestor, a related animal, or than adjacent teeth).

Emigration   departure of animal(s), usually at or about the time of reaching adulthood, from the group or place of birth.

Entelodont   a member of the family Entelodontidae, Oligocene artiodactyls which represent an early branch of the pig family, Suidae.

Epidermis   the outer layer of mammalian skin (and in plants the outer tissue of young stem, leaf or root).

Erectile   capable of being raised to an erect position (erectile mane).

Estrus   the period in the estrous cycle of female mammals at which they are often attractive to males and receptive to mating. The period coincides with the maturation of eggs and ovulation (the release of mature eggs from the ovaries). Animals in estrus are often said to be "on heat" or "in heat." In primates, if the egg is not fertilized the subsequent degeneration of uterine walls (endometrium) leads to menstrual bleeding. In some species ovulation is triggered by copulation and this is called **induced ovulation**, as distinct from spontaneous ovulation.

Facultative   optional (cf OBLIGATE).

Family   a taxonomic division subordinate to an order and superior to a genus.

Feces   excrement from the bowels; colloquially known as droppings or scats.

Feral   living in the wild (of domesticated animals, eg cat, dog).

Fermentation   the decompostition of organic substances by microorganisms. In some mammals, parts of the digestive tract (eg the cecum) may be inhabited by bacteria that break down cellulose and release nutrients.

Fetal development rate   the rate of development, or growth, of unborn young.

Flehmen   German word describing a facial expression in which the lips are pulled back, head often lifted, teeth sometimes clapped rapidly together and nose wrinkled. Often associated with animals (especially males) sniffing scent marks or socially important odors (eg scent of estrous female.) Possibly involved in transmission of odor to JACOBSON'S ORGAN.

Folivore   an animal eating mainly leaves.

Follicle   a small sac, therefore (a) a mass of ovarian cells that produces an ovum, (b) an indentation in the skin from which hair grows.

Forbs   a general term applied to ephemeral or weedy plant species (not grasses). In arid and semi-arid regions they grow abundantly and profusely after rains.

Frugivore   an animal eating mainly fruits.

Gallery forest   luxuriant forest lining the banks of watercourses.

Gamete   a male or female reproductive cell (ovum or spermatozoon).

Generalist   an animal whose life-style does not invole highly specialized

strategems (cf SPECIALIST); for example, feeding on a variety of foods which may require different foraging techniques.

Genus   (plural genera)   a taxonomic division superior to species and subordinate to family.

Gestation   the period of development within the uterus; the process of **delayed implantation** can result in the period of pregnancy being longer than the period during which the embryo is actually developing. (See also IMPLANTATION).

Glands   (marking)   specialized glandular areas of the skin, used in depositing SCENT MARKS.

Graviportal   animals in which the weight is carried by the limbs acting as rigid, extensible struts, powered by extrinsic muscles; eg elephants and rhinos.

Grazer   a herbivore which feeds upon grasses (cf BROWSER).

Grizzled   sprinkled or streaked with gray.

Harem group   a social group consisting of a single adult male, at least two adult females and immature animals: a common pattern of social organization among mammals,

Heath   low-growing shrubs with woody stems and narrow leaves (eg heather), which often predominate on acidic or upland soils.

Herbivore   an animal eating mainly plants or parts of plants.

Heterothermy   a condition in which the internal temperature of the body follows the temperature of the outside environment.

Hindgut fermenter   herbivores among which the bacterial breakdown of plant tissue occurs in the CECUM, rather than in the RUMEN or foregut.

Holarctic realm   a region of the world including North America, Greenland, Europe, and Asia apart from the south west, southeast and India.

Home range   the area in which an animal normally lives (generally excluding rare excursions or migrations), irrespective of whether or not the area is defended from other animals (cf TERRITORY).

Hybrid   the offspring of parents of different species.

Hypothermy   a condition in which internal body temperature is below normal

Hypsodont   high-crowned teeth, which continue to grow when full-sized and whose pulp cavity remains open; typical of herbivorous mammals (cf BRACHYDONT).

Hydracondont   a member of the family Hyracodontidae (order Perissodactyla) which became extinct in the Oligocene.

Implantation   the process whereby the free-floating blastocyst (early embryo) becomes attached to the uterine wall in mammals. At the point of implantation a complex network of blood vessels develops to link mother and embryo (the placenta). In **delayed implantation** the blastocyst remains dormant in the uterus for periods varying, between species, from 12 days to

11 months. Deleyed implantation may be obligatory or faculative and is known for some members of the Carnivora and Pinnipedia and others.

Induced ovulation   see ESTRUS.

Inguinal   pertaining to the groin.

Insectivore   an animal eating mainly arthropods (insects, spiders).

Interdigital   pertaining to between the DIGITS.

Intestinal flora   simple plants (eg bacteria) which live in the intestines, especially the CECUM, of mammals. They produce enzymes which break down the cellulose in the leaves and stems of green plants and convert it to digestible sugars.

Introduced   of a species which has been brought, by man, from lands where it occurs naturally to lands where it has not previously occurred. Some introductions are accidental (eg rats which have traveled unseen on ships), but some are made on purpose for biological control, farming or other ecomonic reasons (eg the Common brush-tail possum, which was introduced to New Zealand from Australia to establish a fur industry).

Ischial   pertaining to the hip.

Jacobson's organ   a structure in a foramen (small opening) in the palate of many vertebrates which appears to be involved in olfactory communication. Molecules of scent may be sampled in these organs.

Juvenile   no longer possessing the characteristics of an infant, but not yet fully adult.

Kopje   a rocky outcrop, typically on otherwise flat plains of African grasslands.

Lactation   (verb: lactate)   the secretion of milk, from MAMMARY GLANDS.

Lamoid   Llama-like; one of the South American CAMELOIDS.

Larynx   dilated region of upper part of windpipe, containing vocal chords Vibration of cords produces vocal sounds.

Latrine   a place where feces are regularly left (often together with other SCENT MARKS); associated with olfactory communication.

Lek   a display ground at which individuals of one sex maintain miniature territories into which they seek to attract potential mates.

Llano   South American semi-arid savanna country, eg of Venezuela.

Loph   a transverse ridge on the crown of molar teeth.

Lophiodont   a member ot the family Lophiodontidae   (order Perissodactyla) which became extinct in the early Tertiary.

Lophodont   molar teeth whose cusps form ridges or LOPHS.

Mamma   (pl. mammae)   **mammary glands**   the milk-secreting organ of female mammals, probably evolved from sweat glands.

Mammal   a member of the CLASS of VERTEBRATE animals having MAMMARY GLANDS which produce milk with which they nurse their young (properly: Mammalia).

Mammilla   (pl. mammillae)   nipple, or teat, on the MAMMA of female mammals; the conduit through which milk is passed from the mother to the young.

Mandible   the lower jaw,

Mangrove forest   tropical forest developed on sheltered muddy shores of deltas and estuaries exposed to tide. Vegetation is almost entirely woody.

Marine   living in the sea.

Masseter   a powerful muscle, subdivided into parts, joining the MANDIBLE to the upper jaw. Used to bring jaws together when chewing.

Melanism   darkness of color due to presence of the black pigment melanin.

Menotyphlan   see Lipotyphlan.

Metabolic rate   the rate at which the chemical processes of the body occur.

Metabolism   the chemical processes occurring within an organism, including the production of PROTEIN from amino acids, the exchange of gasses in respiration, the liberation of energy from foods and innumerable other chemical reactions.

Microhabitat   the particular parts off the habitat that are encountered by an individual in the courese of its activities.

Midden   a dunghill, or site for the regular depostion of feces by mammals.

Migration   movement, usually seasonal, from one region or climate to another for purposes of feeding or breeding.

Monogamy   a mating system in which individuals have only one mate per breeding season.

Monotypic   a genus comprising a single species.

Montane   pertaining to mountainous country.

Montane forest   forest occurring at middle altitudes on the slopes of mountains, below the alpine zone but above the lowland forests.

Morphology   (morphological)   the structure and shape of an organism.

Mutation   a structural change in a gene which can thus give rise to a new heritable characteristic.

Natal range   the home range into which an individual was born (natal = of or from one's birth).

Niche   the role of a species within the community, defined in terms of all aspects of its life-style (eg food, competitors, predators, and other resource requirements).

Nicker   a vocalization of horses, also called neighing.

Noctunal   active at nighttime.

Obligate reqired, binding (cf FACULTATIVE).

Occipital pertaining to the occiput at back of head.

Olfaction, olfactory the olfactory sense is the sense of smell, depending on receptors located in the epithelium (surface membrane) lining the nasal cavity.

Omasum third of the four chambers in the RUMINANT ARTIODACTYL stomach.

Omnivore and animal eating a varied diet including both animal and plant tissue.

Opportunist (of feeding) flexible behaviour of exploiting circumstances to take a wide range of food items; characteristic of many species. See GENERALIST; SPECIALIST.

Order a taxonomic division subordinate to class and superior to family.

Oreodont a member of the family Oreodontidae (order Artiodactyla), which became extinct in the late Tertiary.

Ovulation (verb ovulate) the shedding of mature ova (eggs) from the ovaries where they are produced (see ESTRUS).

Pair-bond an association between a male and female, which lasts from courtship at least until mating is completed, and in some species, until the death of one partner.

Palearctic a geographical region encompassing Europe and Asia north of the Himalayas, and Africa north of the Sahara.

Paleothere a member of the family Paleotheriidae (order Perissodactyla), which became extinct in the early Tertiary.

Palmate palm shaped.

Pampas Argentinian steppe grasslands.

Papilla (plural: papillae) a small nipple-like projection.

Páramo alpine meadow of northern and western South American uplands.

Parturition the process of giving birth (hence *post partum*—after birth).

Pecoran a ruminant of the intra-order Pecora, which is characterized by the presence of horns on the forehead.

Perissodactyl a member of the Perissodactyla (the odd-toed ungulates).

Perineal glands glandular tissue occurring between the anus and genitalia.

Pheromone secretions whose odors act as chemical messengers in animal communication, and which prompt a specific response on behalf of the animal receiving the message (see SCENT MARKING).

Phylogeny a classification or relationship based on the closeness of evolutionary descent.

Phylogenetic (of classification or relationship) based on the closeness of evolutionary descent.

Phylum a taxonomic division comprising a number of classes.

Physiology study of the processes which go on in living organisms.

Pinna (plural: pinnae) the projecting cartilaginous portion of the external ear.

Placenta, placental mammals a structure that connects the fetus and the mother's womb to ensure a supply of nutrients to the fetus and removal of its waste products. Only placental mammals have a well-developed placenta; marsupials have a rudimentary placenta or none and monotremes lay eggs.

Plantigrade way of walking on the soles of the feet, including the heels (cf DIGITIGRADE).

Polyandrous see POLYGYNOUS.

Polyestrous having two or more ESTRUS cycles in one breeding season.

Polygamous a mating system wherein an individual has more than one mate per breeding season.

Polygynous a mating system in which a male mates with several females during one breeding season (as opposed to polyandrous, where one female mates with seferal males).

Polymorphism occurrence of more than one MORPHOLOGICAL form of individual in a population. (See SEXUAL DIMORPHISM).

Polyprodont having more than two well-developed lower incisor teeth (as in bandicoots and carnivorous marsupials). (cf DIPROTODONT).

Population a more or less separate (discrete) group of animals of the same species within a given BIOTIC COMMUNITY.

Post-orbital bar a bony strut behind the eye-socket (orbit) in the skull.

Post-partum estrus ovulation and an increase in the sexual receptivity of female mammals, hours or days after the birth of a litter (see ESTRUS).

Prairie North American steppe grassland between 30° N and 55° N.

Predator an animal which forages for live prey; hence "anti-predator behaviour" describes the evasive action of the prey.

Precocial of young born at a relatively advanced stage of development, requiring a shot period of nursing by parents (see ALTRICIAL).

Preorbital in front of the eye socket.

Preputial pertaining to the prepuce or loose skin covering the penis.

Proboscidean a member of the order of primitive ungulates, Proboscidea.

Proboscis a long flexible snout.

Process (anatomical) an outgrowth or protuberance.

Procumbent (incisors) projecting forward more or less horizontally.

Promiscuous a mating system wherein an individual mates more or less indiscriminately.

Pronking a movement where an animal leaps vertically, on the spot, with all four

feet of the ground. Also called stotting. Typical of antelopes, especially when alarmed.

Protoceratid a member of the family Protoceratidae (order Artiodactyla), which became extinct in the late Tertiary.

Protein a complex organic compound made of amino acids. Many different kinds of proteins are present in the muscles and tissues of all mammals.

Proximal near to the point of attachment or origin, (eg the base of the tail).

Puberty the attainment of sexual maturity. In addition to maturation of the primary sex organs (ovaries, testes), primates may exhibit "secondary sexual characteristics" at puberty. Among higher primates it is usual to find a growth spurt at the time of puberty in males and females.

Puna a treeless tableland or basin of the high Andes.

Pylorus the region of the stomach at its intestinal end, which is closed by the pyloric sphincter.

Quadrate bone at rear of skull which serves as a point of articulation for lower jaw.

Quadrupedal walking on all fours, as opposed to walking on two legs (BIPEDAL) or moving suspended beneath branches in trees (SUSPENSORY MOVEMENT).

Race a taxonomic division subordinate to subspecies but linking populations with similar distinct characteristics.

Radiation see ADAPTIVE RADIATION.

Rain forest tropical and subtropical forest with abundant and year-round rainfall. Typically species rich and diverse.

Range (geographical) area over which an organism is distrubuted.

Receptive state of a female mammal ready to mate or in ESTRUS.

Reduced (anatomical) of relatively small dimension (eg of certain bones, by comparison with those of an ancestor or related animals).

Reingestion process in which food is digested twice, to ensure that the maximum amount of energy is extracted from it. Food may be brought up from the stomach to the mouth for further chewing before reingestion, or an individual may eat its own feces (see REFECTION).

Reproductive rate the rate of production of offspring; the net productive rate amy be defined as the average number of female offspring produced by each female during her entire lifetime.

Resident a mammal which normally inhabits a defined area, whether this is a HOME RANGE or a TERRITORY.

Reticulum second chamber of the RUMINANT ARTIODACTYL four-chambered stomach (see RUMEN, OMASUM, ABOMASUM). The criss-crossed (reticulated) walls give rise to honeycomb tripe.

Rumen first chamber of the RUMINANT ARTIODACTLY four-chabered stomach. In the rumen the food is liquefied, kneaded

by muscular walls and subjected to fermentation by bacteria. The product, cud, is regurgitated for further chewing; when it is swallowed again it bypasses the RUMEN and RETICULUM and enters the OMASUM.

Ruminant a mammal with a specialized digestive system typified by the behavior of chewing the cud. Their stomach is modified so that vegetation is stored, regurgitated for further maceration, then broken down by symbiotic bacteria. The process of rumination is an adaptation to digesting the cellulose walls of plant cells.

Rupicaprid a member of the tribe Rupicaprini, the chamois etc of the Artiodactyla.

Rut a period of sexual excitement; the mating season.

Satellite male an animal excluded from the core of the social system but loosely associated on the periphery, in the sense of being a "hanger-on" or part of the retinue of more dominant individuals.

Scent gland an organ secreting odorous material with communicative properties; see SCENT MARK.

Scent mark a site where the secretions of scent glands, or urine or FECES, are deposited and which has communicative significance. Often left regularly at traditional sites which are also visually conspicuous. Also the "chemical message" left by this means; and (verb) to leave such a deposit.

Sclerophyll forest a general term for the hard-leafed eucalypt forest that covers much of Australia.

Sedentary pertaining to mammals which occupy relatively small home ranges, and exhibiting weak dispersal or migratory tendencies.

Selective pressure a factor affecting the reproductive success of individuals (whose success will depend on their fitness, ie the extent to which they are adapted to thrive under that selective pressure).

Selenodont molar teeth with crescent-shaped cusps.

Septum a partition separating two parts of an organism. The nasal septum consists of a fleshy part separating the nostrils and a vertical, bony plate dividing the nasal cavity.

Serum blood from which corpuscles and clotting agents have been removed; a clear, almost colourless fluid.

Sexual dimorphism a condition in which males and females of a species differ consistently in form, eg size, shape. (See POLYMORPHISM.)

Serology the study of blood sera; investigates ANTIGEN-antibody reactions to elucidate responses to disease organisms and also PHYOGENETIC relationships between species.

Siblings individuals who share one or both parents. An individual's siblings are its brotheres and sisters, regardless of their sex.

Sinus a cavity in bone or tissue.

Sivathere   a member of a giraffe family which became extinct during the last Ice Age.

Solitary   living on its own, as opposed to social or group-living in life-stlye.

Sonar   sound used in connection with Navigation (SOund NAvigation Ranging).

Sounder   the collective term for a group of pigs.

Specialist   an animal whose life-style involves highly specialized stratagems; eg feeding with one technique on a particular food.

Species   a taxonomic division subordinate to genus and superior to subspecies. In general a species is a group of animals similar in structure and which are able to breed and produce viable offspring.

Speciation   the process by which new species arise in evolution. It is widely accepted that it occurs when a single-species population is divided by some geographical barrier.

Sphincter   a ring of smooth muscle around a pouch, rectum or other hollow organ, which can be contracted to narrow or close the entrance to the organ.

Spinifex   a grass which grows in large, distinctive clumps or hummocks in the driest areas of central and Western Australia.

Stotting   see PRONKING.

Subadult   no longer an infant or juvenile but not yet fully adult physically and/or socially.

Subfamily   a division of a FAMILY.

Suborder   a subdivision of an order.

Subspecies   a recognizable subpopulation of a single species, typically with a distinct geographical distribution.

Suid   a member of the family of pigs, Suidae, of the Artiodactyla.

Supra-orbital   pertaining to above the eye (eye-socket or orbit).

Sympatry   a condition in which the geographical ranges of two or more different species overlap (cf ALLOPATRY).

Taiga   northernmost coniferous forest, with open boggy rocky areas in between.

Tarsal   pertaining to the tarsus bones in the ankle, articulating between the tibia and fibia of the leg and the metatarsals of the foot (pes).

Terrestrial   living on land.

Territory   an area defended from intruders by an individual or group. Originally the term was used where ranges were exclusive and obviously defended at their borders. A more general definition of territoriality allows some overlap between neighbours by defining territoriality as a system of spacing wherein home ranges do not overlap randomly—that is, the location of one individual's or group's home range influences those of others.

Testosterone   a male hormone syntheisized in the testes and responsible for the expression of many male characteristics (contrast the female hormone estrogen produced in the ovaries).

Thermoneutral range   the range in outside environmental temperature in which a mammal uses the minimum amount of energy to maintain a constant internal body temperature. The limits to the thermoneutral range are the lower and upper critical temperatures, at which points the mammals must use increasing amounts of energy to maintain a constant body temperature. (cf HETEROTHERMY)

Thermoregulation   the regulation and maintenance of a constant internal body temperature in mammals.

Tooth-comb   a dental modification in which the incisor teeth form a comb-like structure.

Thoracic   pertaining to the thorax, or chest.

Torpor   a temporary physiological state in some mammals, akin to short-term hibernation, in which the body temperature drops and the rate of METABOLISM is reduced. Torpor is an adaptation for reducing energy expenditure in periods of extreme cold or food shortage.

Tylopod   a member of the suborder Tylopoda (order Artiodactyla) which includes camels and llamas.

Umbilicus   naval

Underfur   the thick soft undercoat fur lying beneath the longer and coarser hair (guard hairs).

Understory   the layer of shrubs, herbs and small trees beneath the forest canopy.

Ungulate   a member of the orders Artiodactyla (even-toed ungulates), Perissodactyla (odd-toed ungulates), Proboscidea (elephants), Hyracoidea (hyraxes) and Tubulidentata (aardvark), all of which have their feet modified as hooves of various types (hence the alternative name, hoofed mammals). Most are large and totally herbivorous, eg, deer, cattle, gazelles, horses.

Unguligrade   locomotion on the tips of the "fingers" and "toes," the most distal phalanges. A condition associated with reduction in the number of digits to one or two in the perissodactyls and artiodactyls (cf DIGITIGRADE and PLANTIGRADE).

Vascular   of, or with, vessels which conduct blood and other body fluids.

Vector   an individual or species which transmits a disease.

Velvet   furry skin covering a growing antler.

Ventral   on the lower or bottom side or surface; thus ventral or abdominal glands occur on the underside of the abdomen.

Vertebrate   an animal with a backbone; a division of the phylum Chordata which includes animals with notochords (as distinct from invertebrates.)

Vestigial   a characteristic with little or no contemporary use, but derived from one which was useful and well developed in an ancestral form.

Vibrissae   stiff, coarse hairs richly supplied with nerves, found especially around the snout, and with a sensory (tactile) function.

Vocalization   calls or sounds produced by the vocal cords of a mammal, and uttered through the mouth. Vocalizations differ with the age and sex of mammals but are usually similar within a species.

Withers   ridge between shoulder blades, especially of horses.

Xenarthrales   bony elements between the lumbar vertebrae of XENARTHRAN mammals which provide additional support to the pelvic region for digging, climbing etc.

Yolk sac   a sac, usually containing yolk, which hangs from the ventral surface of the vertebrate fetus. In mammals, the yolk sac contains no yolk, but helps to nourish the embryonic young via a network of blood vessels.

INDEX

Picture Acknowledgements

Key: *t* top. *b* bottom. *c* centre. *l* left.
r right.
Abbreviations: A Ardea. AH Andrew
Henley. AN Nature, Agence
Photographique. BC Bruce Coleman Ltd.
GF George Frame. J Jacana. FL Frank Lane
Agency. PEP Planet Earth Pictures.
SA Survival Anglia.

Cover BC. 1 BC. 2-3 BC. 4-5 BC. 6-7
BC. 8-9 BC. 11 H. Hoeck. 13 WWF/M.
Boulton. 14 W. Ervin, Natural Imagery.
15 NHPA. 16-17 A. 18-19, 18, 19
P. D. Moehlman. 20*t* SA. 20*b* BC.
21 WWF/M. Boulton. 22-23, 24, 25
H. Hoeck. 27 SA. 28-29 BC. 36-37
Musée du Périgord, Périgeux. 38-39 Alan
Hutchison. 39*b* G. Herrmann. 42 M. P.
Kahl. 43 Woodfall Wildlife Pictures.
46*t* SA. 46-47 PEP. 47*t* J. 50 BC.
51*t* M. P. Kahl. 51*b* Woodfall Wildlife
Pictures. 54-55 Nature Photographers.
55*b* WWF. 56 J. 57 GF. 60 Nature
Photographers. 60*b* SA. 61 BC.
62, 64*b* Leonard Lee Rue III. 64-65 AH.
66*b* AN. 66-67 SA. 67*b* R. M. Laws.
68-69 PEP. 70 Nature Photographers.
71 R. M. Laws. 72 AN. 73 AH.
74-75, 75*b* AN. 77 BC. 80 AN.
81 Leonard Lee Rue III. 86 J. MacKinnon.
83 P. Newton. 85 BC. 87 W. Ervin,
Natural Imagery. 90*b* BC. 90-91 J.
91*t*, 92 FL. 93 GF. 94 J. 95 Leonard
Lee Rue III. 96-97 M. P. Kahl. 98 J.
99 R. Pellew. 100-101 M. P. Kahl.
101*t* J. 101, 103 D. Kitchen. 104*b* NHPA.
104*t*, 105 BC. 106-107 PEP.
108, 109 AH. 110-111*t* PEP.
110*c* AH. 110-111*b* PEP. 114, 115
D. Lott. 118, 119 A. Bannister. 120 PEP.
121 P. Wirtz. 124-125 A. Bannister.
125*b* Leonard Lee Rue III.
126-127, 128 PEP. 128-129 Eric and
David Hosking. 129 SA. 132, 133
M. Stanley Price. 134-135 A. Bannister.
135*b* W. Ervin, Natural Imagery.
138-139 PEP. 138*b* J. 139 W. Ervin,
Natural Imagery. 139*r* GF.
142 M. P. Kahl. 143 W. Ervin, Natural
Imagery. 146-147 NHPA. 147*b* Leonard
Lee Rue III. 150-151 BC. 151*b* G. di Nunzio.

Artwork

All artwork © Priscilla Barrett unless stated
otherwise below.
Abbreviations: SD Simon Driver. ML
Michael Long. AEM Anne-Elise Martin.

10 ML. 12 Malcolm McGregor. 14 SD.
30, 31 SD. 32*t* ML. 32 SD. 33 ML.
34, 35, 36 ML. 41*t* AEM. 41 ML.
59*t* AEM. 59 ML, SD. 70, 80, 93 SD.